LIQUID

How CEOs & CTOs Unlock
Flow and Momentum in
Complex Systems

Kathy Keating
Etienne de Bruin
Scott Graves

**CTO
SENTINEL**

CTO
SENTINEL

LIQUID

How CEOs & CTOs Unlock Flow and Momentum in Complex Systems

FIRST EDITION

Published by CTO Sentinel
ISBN 978-1-967830-00-8 *Hardcover*
978-1-967830-01-5 *Paperback*
978-1-967830-02-2 *Ebook*

For more information about the book or to contact the author, visit www.ctosentinel.com.

Dedicated to CTOs far and wide.

May this book be an oasis on your journey.

Table of Contents

Preface

When Scott, Kathy, and I headed to the Colorado mountains to work on our book, we had no idea how we'd do it. We'd just started our little company, LevelsOS, LLC, because we knew that we'd discovered something beautiful and we wanted to work on it together.

We rolled into a tiny ski town called Winter Park, where Kathy had booked us an Airbnb in which we planned to stay four days to work on our book.

How do three people who have never co-authored a book together find a rhythm that feels generative and optimistic? I'm sure these were the questions floating through my colleagues' minds, as they were very much alive in mine. But what ensued was nothing short of magical.

We're three very different people who somehow found a way to fold our strongest convictions into wonderful conversations that led to pages and pages being written before we knew it.

I think the most special moment in our writing process was when Kathy suggested we bring our ideas to life through a fictional founder. That's how Alice was born. She became the central character in the story that weaves through this book, helping us illustrate the real challenges and decisions technology leaders face as their companies grow. Soon after, Theo, her trusted technical partner, joined the journey.

We all felt instantly connected to Alice's vision and shared in her excitement as she and Theo set out to build their startup. And yes, we laughed often, imagining all the classic mistakes they were bound to make along the way.

Capturing Alice's story as a way to express our collective experience was one of the most important exercises I could have ever undertaken. It's a story in which I hope you will find yourself as well. And if her story isn't where you're at in life, then that's okay also!

You'll find the story of Alice and Theo's journey—from building their startup to scaling it into an enterprise—throughout the book as a way to ground the concepts in real-world situations. Their experiences will help you connect systems thinking to the day-to-day challenges you face, like managing team dynamics, scaling operations, aligning priorities, and navigating the complexity that comes with growth.

I know that there was a moment in time when I drew a grid on a sheet of paper all by myself, and I called it CTO Levels. But it was never just my moment. My pen was playing its part in a self-organizing symphony, and Kathy and Scott were both seated right next to me.

I hope you'll enjoy reading our book. And we hope that you'll enjoy being a passenger on our journey.

But most of all, I hope this will be the beginning of our collective journey as we seek to build better tech companies with exponential impact toward the greater good.

– Etienne de Bruin

This is Just the Beginning

Others are already applying what's inside, and you can too.

Scan the QR code or visit ctosentinel.com/liquid to unlock bonus

tools, frameworks, and content to help you put Liquid into action as

you build a more adaptive organization.

Liquid

PART 1:

The Secret World

Do you believe that my being stronger or faster has anything to do with my muscles

in this place? Do you think that's air you're breathing now?

–Morpheus to Neo, *The Matrix*

Liquid

Chapter 1:
The Secret Life of
Software Companies

Neo, the protagonist of *The Matrix*, a popular 1999 action sci-fi, suspected that the world was not as it seemed. He felt trapped, working hard but never truly getting ahead. It was as if something vital, some deeper understanding, was just beyond his reach, eluding him at every turn.

Unbeknownst to Neo, his life was governed by a hidden system, an invisible world influencing his every move. This secret world controlled his reality, shaping his destiny in ways he could neither see nor understand. His journey began when he chose to uncover this truth, stepping into the unknown to confront the system and learn how to navigate it.

Many of us in the tech world feel the same way. We operate in environments that appear chaotic, we feel we can never get a handle on why things aren't working the way we need them to, and we rarely feel like we're getting ahead. We struggle to see the connections and hidden structures that dictate success or failure. Success feels like it's always just out of reach.

There's a secret world determining the outcome of your technology team and your company—and you just cannot see it. This book is about unveiling this hidden reality, helping you understand the complex systems you navigate daily and giving you the skills to confidently see the path toward success. Once you can see the system at work within your teams, you can influence it.

Neo had to make a hard choice and take a big leap to see the world as it really is. Similarly, as technology executives, we must make bold decisions and embrace new perspectives to understand and harness the complex adaptive systems (CAS) within our organizations. Only by doing so can we unlock our true potential and lead our teams toward success.

Are you ready to open your eyes and peek behind the veil? If you trust us enough to take this journey, we'll reveal the hidden system quietly shaping your technology organization—the same system that determines whether your team moves with speed or stalls in complexity, and whether your products scale elegantly or collapse under their own weight.

You'll learn to see the patterns that govern the flow of work, the silent forces that influence collaboration and decision-making, and the structures that either accelerate or constrain your company's growth.

By the end, you'll no longer be guessing why things feel stuck or out of control. Instead, you'll understand how to lead your organization with clarity, influence outcomes with precision, and build a business that thrives in the face of complexity.

Come with us and discover just how deep the rabbit hole goes.

Pulling Back the Veil

True success lies in acknowledging and embracing the complexity of the system in which we operate. Yet research[1] shows that many leaders struggle to recognize the complexity of the systems within which they operate.

The complexity of modern businesses lies in their interconnected and dynamic nature. Organizations are made up of numerous components (people, processes, teams, and technologies) that interact in ways that aren't always straightforward or predictable. These interactions often result in unexpected outcomes, making it difficult to see the whole picture. As a result, many leaders develop oversimplified views of how their businesses operate, overlooking the intricate dynamics that drive success or create challenges.

In an attempt to bring some semblance of control to their business, many leaders apply simplistic decision-making models to address their complex problems. They fail to acknowledge the nuances and interdependencies within their organizations. Complex challenges are rarely solved with a simple solution.

If simple solutions could solve complex challenges, then 50% of businesses wouldn't fail in their first five years.

Traditional leadership models often emphasize linear thinking and control-based approaches, which can be insufficient in high-growth companies where unpredictability and emergent behaviors are prevalent. This misalignment can lead to ineffective decision-making and an inability to address multifaceted challenges effectively.

To address this, the Systems Thinking Alliance[2] emphasizes that systems leadership requires a fundamental shift away from control-based management and toward dynamic engagement with others. In complex environments, leaders cannot solve problems in isolation. Instead, they must engage in collective and systemic processes of change, recognizing their role within a larger ecosystem that includes multiple stakeholders with diverse needs and perspectives.

Similarly, Uhl-Bien et al. introduced complexity leadership theory (CLT)[1] as a framework to better understand the shifting roles and actions of leaders in CAS. This framework emphasizes the need for leaders to enable learning, creativity, and adaptability within their organizations, moving beyond traditional bureaucratic functions to embrace emergent, informal dynamics.

These insights suggest that to navigate complex systems effectively, leaders must adopt a systemic mindset, fostering collaboration and co-creation among diverse stakeholders. By doing so, they can better understand and influence the intricate networks and relationships that characterize complex organizational environments.

Leaders must move beyond surface-level metrics and develop a deeper appreciation for the intricate dynamics at play. When leaders learn to recognize the complexity at work within their business and build the skills necessary to influence change in their desired direction, they will achieve their desired results.

True success lies in acknowledging and embracing the complexity of the environment around us, which requires adaptability and continuous improvement.

Building Versus Growing

Great software and great products aren't built, they're grown. And the quality of the harvest depends on the quality of the soil in which it's planted. Who walks into a storehouse, sees bags of corn, and thinks that somebody made them?

A farmer doesn't make corn; a farmer grows corn. The farmer's role is to establish conditions for growth. By preparing the soil, planting seeds, and providing water, the farmer facilitates the natural processes that allow corn to grow. Similarly, leaders must cultivate an environment that enables innovation and adaptation. We cannot control every outcome directly, but we can significantly influence success by nurturing the right conditions—much like a farmer ensures the health and growth of their crops through careful management and support.

Let's consider a different example that also highlights how systems work. Think about the refrigerator that dumps ice into your cup. Where did that ice come from? It came from a delicate ecosystem where a complex system of components (compressor, evaporator coils, water lines, and thermostat) maintains the precise conditions for freezing water into shaped containers. We can rarely directly produce outcomes. Instead, we can only orchestrate and influence the ecosystem by ensuring all parts work together harmoniously, cultivating an environment where desired results naturally emerge. Understanding and managing these underlying systems is crucial for achieving long-term success and adaptability.

Are You Willing?

This book brings the secret world of complex systems, like your business, into the open. No longer will you be at the mercy of something you don't understand or recognize. You may have thought of your business operations as a black box, expecting your team to perform miracles because they should "know how to do their jobs," like squeezing water out of an empty sponge. But it's time to stop wringing that dry sponge over and over again expecting different results.

This secret world isn't made of code, requirements, or budgets. It's composed of boundaries, interactions, and adaptations. Its emergent qualities can't be directly controlled or manipulated but instead require nurturing through consistent influence.

Most leaders believe that driving their business forward is about managing tasks: completing projects, hitting deadlines, and staying within budgets. While these are important, they're not the full picture. A business isn't simply the sum of its tasks. It's the intricate system that governs how those tasks are executed, who executes them, and how they interact with one another.

To illustrate this idea, imagine a finely tuned orchestra. The individual musicians (tasks) play their instruments, but the harmony (the system) emerges only when the conductor ensures that all elements work together in balance and rhythm. Without the system guiding these interactions, you'd be left with disjointed noise, not music.

Similarly, in your business, the tasks are just the surface-level activities. The true force driving success is how those tasks are organized, how boundaries are defined, and how you adapt when circumstances inevitably change.

Focusing solely on tasks is like listening to one musician in an orchestra and expecting to hear the full composition. You might catch a few beautiful notes, but without the conductor bringing the entire ensemble into balance, the music never fully comes to life. In the same way, your business depends not just on individual tasks being completed but on how those tasks are coordinated, timed, and shaped by leadership to create a cohesive whole.

The conductor doesn't play every instrument, but they ensure each section enters at the right moment, blends with the others, and serves the integrity of the piece. As leaders, we play the same role: guiding the flow of work, defining clear boundaries, and ensuring the entire system moves in rhythm toward shared outcomes. Tasks are simply the visible outputs of your system—whether that system is creating a smooth, coordinated harmony or unraveling into chaos, inefficiency, and rigidity.

Consider a product development team that delivers features late or riddled with bugs. At first glance, the issue might appear to be with the developers themselves, but the root cause could lie in the broader system. Perhaps the engineering team isn't receiving clear priorities from product management, or maybe their workflows are bogged down by outdated tools. Similarly, a sales team struggling to meet targets might seem like they need more training or motivation, but the real issue could be misaligned incentives, a lack of integration with marketing, or inadequate CRM tools creating bottlenecks. Technology, too, plays a critical role. An underperforming infrastructure might slow down a website, frustrating customers and increasing churn, but the surface symptom could simply appear as a dip in sales.

To lead effectively, you must look beyond individual tasks and outcomes to understand the system as a whole. How do signals—priorities, feedback, or updates—flow through your organization? Are there clear boundaries that allow teams to focus on their core responsibilities, or is everyone constantly stepping on each other's toes? How resilient is your system when unexpected challenges arise? Leadership requires not just managing tasks but designing and influencing a system that supports sustainable success.

Business isn't just about going through the motions to achieve all of the tasks, blindly believing that success will follow. Successful business leaders understand how to wield the invisible forces that can enable or hinder what gets done.

If you're willing to come with us on this journey to understand the intricate system dynamics at play, you'll learn how CTOs can create solutions that not only meet immediate deadlines and estimates but also ensure low defect rates, fast changes, and predictable costs.

Through this journey, you'll find your way to becoming a grower of systems that produce the technologies that drive your company's success. Just like Neo in *The Matrix*, seeing the system operating around us is the first step. Let's explore how we, as your authors, each found our way out of the Matrix and learned to see the system clearly.

Chapter 2:
Etienne's Story—How the Sentinel Came to Be

When the pandemic hit the planet, all hell broke loose in the tech industry. Developers, managers, and executives were all sent home. No more commutes, no more unnecessary contact, and a lot of "work from home" conversations. Companies were adopting crisis management communication protocols, startups were wrapped in uncertainty, and we all hunkered down in our homes, awaiting a COVID-19 winter.

This was when I found myself contemplating my own future. I was the founder of 7CTOs,[3] a premium membership community coaching Chief Technology Officers. Our revenues depended on the professional development budgets of technology companies around the world. The same world that had just shut down. And with that, professional development budgets evaporated.

I was afraid. We'd been cruising at 7CTOs with year-over-year growth. I'd just appointed a CEO to help me take 7CTOs to the next level. But instead, my new CEO was going to have to work on holding our member base steady at the current level. Pretty soon, membership cancellations started rolling in. It wasn't a mass exodus of members. It also wasn't a sudden onslaught.

There was a gradual slowdown in sign-ups and a gradual increase in cancellations. For an in-person, events-driven company, 7CTOs was facing a very uncertain future. And so was I. It was time for me to do something I hadn't done since founding 7CTOs in 2013: go back into the job market and find myself a CTO job.

Over the years, I'd established a solid network of founders because I'd been talking to them about enrolling their CTOs in 7CTOs. So, I went to work, putting my name out there as a CTO for hire. Soon I had a shortlist of companies that were in need of a CTO. Before too long, I'd landed a gig.

I became the CTO of ACE, a SaaS company in a niche video-streaming space with great founders and a solid plan for growth. I was very excited. I finally had the opportunity to be a CTO again and apply all the coaching knowledge I'd accumulated over the years into the real-world needs of a scaling startup.

It was a disaster.

A Square Peg

I walked into a chaotic situation at ACE. Everything was on fire. The team was always one code commit away from the whole house burning down. The founders had been seduced by their ambition to become the market leaders.

This meant there was a constant flow of feature requests. The departing CTO had a proverbial suitcase filled with secrets that no one had realized they should have asked for when he left.

And all eyes were on me to bring order to this chaos. I did have my plans. Plans to save the company from itself. Armed with models and playbooks we'd developed through the years, I felt that a systematic approach would bring some calm and clarity.

I'd developed a framework called "CTO Quadrant" that neatly packaged the four things a CTO should dominate within a startup in order to ensure technology success:

1. Technology
2. Team
3. Tools
4. Timing

Four Ts! Four Ts that looked gorgeous in a YouTube video. In fact, to this day it's one of the most viewed videos when searching for "What is the role of a CTO?" The only problem was that implementing the CTO Quadrant was like hammering a square peg into a round hole.

I diligently focused on various aspects of our technology (Technology). I confidently led the team, nurturing them while also challenging them to meet higher standards (Team). I researched tools in light of immediate cost savings but also for future scale (Tools). And I was in regular conversations with the founders on the timing of what we had to focus on (Timing).

But it didn't land with the founders. Our meetings devolved into how long it would take to get features critical to expansion revenue released. We were conversing at the tactical level when the business needed us to be on the same page about sustainable business growth. The founders were rightfully concerned about growing costs and what that meant for budgets in the near future. They were terrified about not maintaining their competitive edge. And, of course, they needed to see revenues increase.

No matter how much CTO-ing I was doing, it felt like a giant waste of time. I was trying to be pragmatic about helping the organization, but my work went unnoticed as it was eclipsed by the blind determination of the C-suite to release more features. We weren't getting anywhere.

After a slow burn of frustration and concern for what I'd gotten myself into, I grabbed a blank piece of paper one afternoon. I allowed myself to pierce through the veil of my ego and expertise to allow the truth of my struggle to flow freely through my consciousness. Why was this so hard? What was I missing in helping the CEO understand the impact of my work? And then the floodgates opened.

It dawned on me that my focus all these years had been on *the work* of the CTO. To firstly be methodical about the technology that needed to be built and secondly to be courageous about delivering that technology to the organization. I had blind faith that doing *the work* would be enough to serve the business, at which point the business could take care of itself.

But what if the business couldn't take care of itself? What was the point of making a success of the technology if the business wasn't going to be able to ingest that value and translate it to its customers? I realized that my CEO and I were like two ships passing each other in the night. He was focused on business first as the CEO, and I was focused on technology first as the CTO.

What if I removed *technology first* from the equation? What if my role as CTO was *technology second*? What if I put the needs of the business before the needs of technology?

It was then that my pen started scribbling on the blank piece of paper in front of me, like when Jackson Pollock first dripped paint on to his blank canvas.

The Birth of the Sentinel

As I reflected on my role at ACE, I asked myself, "What effect do I have on the business when I do an outstanding job as a CTO?" The first word that came to mind was *speed*. Undoubtedly, if I did my job well as a CTO, the organization would feel like technology was delivered speedily. Expeditiously. With exceptional quality and precision. Targeted beautifully toward the needs of its customers with minimal waste.

Secondly, I thought of how dependent the founders were on my ability to shape the organization. Not only to project costs, but also to clearly map the business's future needs to today's planning. I thought about how hard that conversation could be and how it required courage and conviction from the CTO. To me, this was embodied in the word *stretch*: a determined action that causes discomfort for the existing organization.

The third word that basically wrote itself was *shield*. Clearly, a company looks to the CTO to shield it from current and future threats—not only the actors that might cause harm but also the technology stack that could crumble and fall. Lastly, I realized that unless the CTO is aligned with the rest of the C-suite in chasing revenue, they don't deserve a seat at the leadership table. This meant not only having a revenue focus but also being influential inside the organization, with an ability to sell. Sell ideas. Sell vision. Sell the product.

And so my fourth word was *sales*.

Four Ss!

1. Speed
2. Stretch
3. Shield
4. Sales

With these four outcomes in place, I began to see the role of the CTO differently. It was no longer just a collection of responsibilities or technical objectives. The CTO was a sentinel role. A watch standing guard through the night, quietly protecting the organization from its own potential downfall.

For the first time in my career, I felt like I had a clear, outcomes-based approach to what it truly meant to be a CTO. It was no longer about looking inward, focused solely on the technical work, but about turning outward toward what was best for the business as a whole. I realized I didn't want to be in the ivory tower anymore.

I didn't want to lead with technology first, as if that alone could ensure success. What mattered was putting technology in service to the livelihood and sustainability of the company. I saw the CTO role as one that focused on delivering speed, maturing the organization as it grew, shielding it from future risks, and being a voice of reason and influence within the leadership team. That was the work I wanted to do and the sentinel I aspired to become. The CTO Sentinel.

I Quit

However, despite this new understanding, reality hit hard, and it spilled over into my CTO gig at ACE. Well, we had a disastrous launch of their new product, and I took responsibility for my part in the flub. I recruited their old CTO back to the organization, and I quit. The COVID-19 pandemic was underway and a global economic contraction was just beginning.

I felt terrible. A very difficult nine months had led to an upgrade in my understanding of the role of CTO, but it had also made me realize that I'd failed miserably at giving ACE the CTO they needed.

Introspection

Shortly after this experience, I went into a mild depression. I spent days introspecting and scrutinizing my performance as a CTO over the years. I'd successfully exited a company that I'd co-founded a few years prior as CTO.

I'd started a CTO community called 7CTOs that was very successful. I'd held numerous advisory board roles as the technical representative. Yet, I'd failed as CTO at ACE.

I started looking at the aspects I'd missed walking into the role at ACE. The CEO was urgently pushing for features. The existing CTO had fallen out of favor with the C-suite. The codebase was a house of cards, and the company was on the verge of committing resources and time to creating their own video-streaming service in support of their real business.

It struck me that I'd accepted the gig based on a false premise. The CEO needed me to fix urgent issues. We were both, however, unaware that these issues were merely symptoms of a deep-rooted problem: The complexity of the software had outpaced the team's ability to manage it, much less build on it. No amount of pushing, hiring, or firing was going to produce the results the business needed.

Looking back, I often ask myself: What could have been done differently to better prepare me—or anyone—for the role of CTO at ACE? The truth is, the leadership team needed a clearer lens through which to understand the complexity of the entire system, not just its symptoms. If we'd had a framework grounded in systems thinking, we could have diagnosed the real challenges the company was facing.

We would have seen that the problems weren't just technical—they were systemic, intertwined with the organization's structure, priorities, and pace of growth. And perhaps, with that understanding, we would have realized that I wasn't the right fit for what ACE needed at the time. Or maybe we would have adjusted our expectations about what any CTO—no matter how qualified—could realistically accomplish in that situation. Most importantly, the CEO would have had better insights into the root issues holding the company back, instead of hoping that a new CTO alone could be the hero who turned things around. That's the real question: not who the right person was for the role, but whether we were looking at the right system in the first place. And without a systems thinking lens, it's almost impossible to answer that.

The CTO Pyramid

Instead of looking at a company through the lofty goals it had for its technology products and services, I started looking at it through the lens of the business of being a business. Ultimately, for the CTO, a business consists of a simple idea: Set a financial target, and establish a budget to hire and nurture a team that builds the technology that helps the company reach that financial target.

What if the technology challenges a company had was a lagging indicator of the business challenges it had? I drew a triangle, and at the top I placed the word "Revenue." Beneath it, I created the second level to the triangle, and I wrote "Budget." Next to it, I wrote "Tech." Very simple. The CTO Pyramid. What I didn't realize is that this pyramid would lead me to a very special discovery.

CTO Levels

If we consider that regardless of company size, complexity, stage, or influence, the CTO Pyramid stays the same, then we can begin to ask ourselves what a company's technology needs are at various budget and team sizes. I grabbed a pen and paper and drew a grid. On the y-axis, I wrote down some budgets and corresponding development team sizes. I started with a team of one and imagined a negligible budget as a starting point. This would represent what most startups face. Then I enumerated all the way up to a team size of 400 and a budget of $50m. This became what we now know as CTO Levels.[4]

Chapter 3:
Scott's Story—Why Am I Struggling to Scale?

It is interesting to contemplate an entangled bank, clothed with many plants of many kinds, with birds singing on the bushes, with various insects flitting about, and with worms crawling through the damp earth, and to reflect that these elaborately constructed forms, so different from each other, and dependent on each other in so complex a manner, have all been produced by laws acting around us.
–Charles Darwin, *On the Origin of Species*[5]

My name is Scott Graves. I'm a CTO and consultant who specializes in scaling technology organizations. Like many CTOs, my journey to the C-suite took place in an engineering organization. I started as a journeyman software engineer and worked my way up.

My experience as an engineer and leader prepared me for many of the problems I encountered in technology departments. I learned to architect for the future. I learned to manage technical debt. I learned to coach developers and team leads. But it didn't prepare me for how radically my job changed as my department scaled up. The first departments I led had scaled slowly, a circumstance I could handle by trial and error. But, in 2018, my luck ran out. I was the CTO of a technology startup that algorithmic-traded certain volatile assets. This company went from $0 to around $300m in volume in two years.

I found myself struggling to meet the challenges of managing my rapidly growing department. Collaboration in the C-suite became acrimonious. The product operation I stood up couldn't identify fruitful lines of investigation. My machine learning operation stopped making progress. It was taking longer and longer to ship features. I didn't understand why.

One day, the young CEO of the company came into my office and asked me a provocative question: "How good of a CTO are you?" I was taken aback and slightly offended. "What do you mean? Look at how well we're doing—I think I'm doing pretty well." But he pressed the issue: "I don't mean how well are you doing here. I mean on an absolute scale from the best CTO to the worst, how good are you?"

I couldn't answer his question. There was no absolute scale of CTO quality that I knew of or could find. My CEO said: "I want you to find a coach. It doesn't matter how good you are; even LeBron James has a coach." My pride was hurt, but I couldn't deny that he had a point. Also, I knew that things were spinning out of control and I needed advice. That's how I met Etienne de Bruin.

I joined 7CTOs in late 2018 and was introduced to a great group of CTOs who helped me overcome many of the tactical problems I was facing. Their insights made an immediate difference. I applied what I learned and saw progress. We shipped features faster, improved cross-team collaboration, and made key hires that filled critical gaps. For a time, things felt like they were moving in the right direction. Even so, something was missing. Despite all the tactics and tricks I picked up from my colleagues, I couldn't shake the feeling that I wasn't progressing on my CEO's hypothetical CTO excellence scale. I didn't understand why I did so well in small organizations but struggled when they grew past a certain size. The more I thought about it, the more I realized I was searching for something deeper than another tactic. I needed to understand the system at play.

In 2020, the global COVID-19 pandemic disrupted my CEO's company. The assets we traded became unavailable worldwide and our operations ground to a halt. While I waited for the pandemic to end, I took part-time CTO work at another company. I applied what I'd learned from 7CTOs and things went great. We had solid processes, delivered reliably, and achieved positive results. Still, I found myself coming back to the same question: Why had I struggled so much in my previous role? There had to be more than a collection of best practices. Was it possible that there were deeper principles that governed the growth of technology organizations?

This question led me to found ScaleTech Consulting.[6] I wanted to systematize everything I'd learned, find patterns that explained the challenges technology organizations face as they scale, and use that knowledge to help companies grow smoothly. I built a framework called "Transform" that aimed to do just that.

Transform was able to explain some of the patterns in my clients' challenges; still, I wanted something deeper—a theory that could be applied more broadly, not just to specific situations. I was searching for a model that could stand the test of scale and complexity, one that could offer clarity no matter the size of the organization. Little did I know, my friend Etienne was on the same journey.

One day, he scheduled a call with me to show me something he was working on. It was an 8×10 piece of paper covered with a grid of squares with tiny text. This was the seed that would become CTO Levels. As he walked me through his creation, I realized that this crowded sheet of paper contained Transform inside it. But it also had two things that made it a general theory.

Those two things were the Sentinel and the Levels. The Sentinel describes the emerging properties by which we can measure a healthy technology department as it grows. The Levels act as a ruler that measures the complexity of the challenges within your technology department. The squares ("Blocks") tell you what a CTO should be doing for each of their responsibilities at each level of complexity. I could see that Transform was confined to the blocks in the first three levels. And I could see that there was much more above those foundational levels.

When the call ended, I abandoned Transform and resolved to work with Etienne on CTO Levels. I could see the shapes and potentialities swimming beneath the theory. Thanks to the Levels scale, the scales had fallen from my eyes.

Etienne's little sheet of paper illustrated to me that the way a CTO's job changes isn't a smooth climb up a mountain of complexity. There were sharp drops and nearly vertical walls. The Levels scale was discontinuous and spiked. Between Levels 3 and 4 was a sharp uptick in complexity. Etienne and I would later call this sheer climb the "Level 4 Organizational Complexity Wall." This was what was causing me so many problems in 2017. I'd found the beginning of an answer to that provocative question: "How good of a CTO are you?" I was a Level 3 CTO.

I hadn't yet learned to climb the wall at Level 4, but I would. Along the way, I would gain many more insights into the wonderful machines that CTOs and their teams build called "technology departments."

Together, Etienne and I have discovered other scales and other kinds of complexity. These scales have refined our understanding of what the Levels are and what they mean. Like Darwin, we find ourselves contemplating our own entangled bank—one made not of plants, birds, and worms but of people, processes, and technology. We were close to a breakthrough. But before that could happen, we needed one more teammate.

Liquid

Chapter 4:
Kathy's Story—Success Requires Partnership and Teamwork

My software engineering journey began in the most unexpected way. In my first role as a junior engineer at Digital Equipment Corporation's AI Research Center, I was building commercial artificial intelligence (AI) before people could even spell AI. I traveled the globe to interview mobile telecom CTOs in order to figure out how to apply AI to address rampant mobile fraud. As a deeply introverted person, I was absolutely terrified and had no idea what I was doing.

However, our amazing team was united toward delivering on this mission, and our results were unprecedented—our product dominated the worldwide market and still exists decades later. Despite my deeply rooted fears, this early experience taught me the importance of collaborative teamwork and working toward a unified mission as a team when navigating uncharted challenges. As a junior engineer, I remember marveling at the adventure I'd had with this incredible team and eagerly anticipating how the rest of my career would unfold in a similar manner. And it did, for a while....

As my career as a software engineer evolved, the birth of the internet opened up even more exciting adventures with dynamic and effective teams. I joined a startup that built one of the first internet content management systems. I acted as the conduit between our engineering team and the clients, helping them understand the internet and implement it successfully for their businesses.

When the dot-com bubble burst in 2001, our company was acquired along with several others, but we were the only one that didn't eventually go under. That product's survival and continued success reminds me how important a strong foundation and a highly aligned team are as part of an effective recipe for success.

I soon landed at a fast-growing startup called Monster, where I led the employer engineering teams. The rapid growth was exhilarating but came with many challenges—our tech architecture was brittle and complicated. It was difficult to decipher what marketing wanted us to build next. Out of sheer necessity to try to move my team forward, I began translating marketing requests into product requirements.

Through trial and error, I guided my teams to deliver faster, deploy more effectively, and build stronger features. My efforts were noticed, and I was eventually tasked with founding the product strategy team, creating the first clear road map, and experimenting with implementing the tenets of the Agile Manifesto into our team's flow.

Halfway through my career, everything was going amazingly! Then I hit a brick wall. In 2005, I moved from the thriving tech hub of Massachusetts to the nascent tech scene in Colorado. Suddenly, I found myself working with executives who dismissed merit and skill in favor of their networks and egos. WHAM! That hurt. I soon learned, through a series of agonizing roles, that while someone might have the skills and insight to drive growth, they can only be activated within a system that supports it. An arrogant new CTO at one company claimed there wasn't a single good software engineer in Colorado because they all lived in Silicon Valley.

Consequently, 80% of his engineering team, including me, eventually walked out on him. Another CEO repeatedly told me, "I don't care what you say; just do what I tell you to do," like a broken record, week after week. This was an incredibly stressful period in my career. It was gut-wrenching to watch that company slowly fail, knowing that despite having many effective tools I could wield, my voice wasn't being heard and thus I couldn't make a difference.

With my hands tied behind my back, all I could do was simply watch these companies flounder and fail. It was at this moment of my career that I committed to not watching the car wreck happen in front of me. I was going to continuously grow into the most effective technology leader I could be. And I would attempt to partner only with those CEOs who respected and valued my partnership and who also wanted to learn and grow. We would do it together.

It was during this time in my career that I began to truly understand the opportunities for failure in the complex system around me. I became aware of what could make the system flow smoothly and what could cause it to grind to a halt. I started seeking what I called "flow state" and practiced maintaining it as much as possible in my role, because that's where I knew the magic continued to happen.

What I came to understand as "flow state" was more than just a feeling of momentum or ease—it was the signal that our system was working.

For me, flow state looks like this:

- The company and our people are aligned, top to bottom, with our mission and goals.

- Our processes are in harmony with our needs—structured enough to create clarity yet flexible enough to adapt.

- We deliver value consistently across the business, without grinding gears or wasting effort.

- We say what needs to be said, with honesty and care, knowing we're all working toward the same outcome: helping the company win.

- Our collective efforts are focused on where the business needs us next, not just what's immediately in front of us.

- We proactively steer what's coming, rather than constantly reacting to what's happening.

- We can see when we're in flow and when we're not.

- No egos. No power plays. Just partners, working together across functions toward a shared purpose.

At the time, I didn't realize it, but what I was describing wasn't just a state of flow. It was systems thinking in practice. What I was really learning was how to step back, see the business as a complete system, and find ways to keep everything working together and moving in the right direction. Flow state was simply the experience of a healthy system in motion. The opportunity to truly put flow state into action came when I joined a FinTech company led by two bootstrapped founders, post-financial recovery, in 2015.

They trusted me to build a new business line from scratch, and we went from concept to launch in just three months. Within a year, we became a top lender in our space. This whirlwind period of constant movement, pivoting, and refining allowed me to fully embody systems thinking, operating in a highly effective flow state almost every day. Seeing this product in the wild today instantly reconnects me to who I want to be in this world.

I went on to found my second company, Apostrophe, an innovative HealthTech startup, in 2016. I naturally fell into my flow state and applied it to everything I did. We thrived, achieved early revenue, went through the Techstars Boulder accelerator, secured venture capital (VC) investment, and grew rapidly. Apostrophe was ultimately acquired in 2021.

Leading a fast-growing company is intensely challenging. Maintaining flow state required us to move quickly (speed), constantly level ourselves up (stretch), handle health data responsibly (shield), and work effectively with others (sales). This experience forced me to stay humble, embrace failures, deal with complex founder dynamics, be introspective, and stay focused in the face of what seemed impossible.

It was during this period of founding Apostrophe that I met Etienne de Bruin and joined 7CTOs. Little did we know then that my flow state and his Sentinels were so closely aligned.

When he shared his CTO Levels framework with me, I realized that I'd found a like-minded soul. Another CTO saw our role the same way I did! I wasn't alone. More recently in my career, I've had the privilege of leading transformational changes for businesses across the spectrum of the CTO Levels framework, guiding teams from early-stage startups (Level 0) to highly scaled organizations (Level 10). The fear and uncertainty I felt as a junior engineer have long since evolved into a deep respect for the complexity of organizations and the courage it takes to lead them successfully.

Today, as a CTO Executive Coach, Strategic Advisor, and occasional Interim CTO, in my work through Kathy Keating Consulting[7] I often step into challenging situations, many times after a previous CTO has been let go. My role is to help struggling product and engineering teams find their way back to a flow state where they can operate with clarity, alignment, and confidence. But my true mission goes beyond turning around product engineering operations.

My greatest passion is empowering CTOs to navigate their roles with insight and intention, to help executives see the system at work within their companies, and to ensure they never reach a point where their value is questioned.

By helping leaders grow and adapt, I aim to create lasting impact, not just for the companies I serve but for the people within them. Because when a CTO thrives, their teams thrive, and the ripple effects of that success can transform entire organizations.

Chapter 5:
Alice Starts Her Company

Have you ever had a dream, Neo, that you were so sure was real? What if you were unable to wake from that dream? How would you know the difference between the dream world and the real world?

–Morpheus, *The Matrix*

Humans naturally crave simplicity. We want clear answers, straightforward paths, and tidy solutions. Yet beneath the surface of every company lies a world that defies such neatness: intricate systems, unseen connections, and emergent behaviors that shape outcomes in unpredictable ways. What if the key to success wasn't in simplifying this complexity but in learning to embrace and influence it?

To make these concepts more relatable and easier to grasp, we'll utilize the story of a fictional company throughout this book. Alice is our founder CEO, and Theo is her first technology hire as the CTO. This company, though fictional, is deeply rooted in the reality of our own experiences and the common challenges we've often faced.

By weaving Alice's story throughout the book, we aim to provide concrete examples and scenarios that mirror the real-world challenges you face. This will help you recognize and navigate the dynamic interactions within your own company that directly influence your success. As we follow Alice's journey, we'll explore the underlying systems that shape the development of her innovative product— systems that also exist within your business. You'll see how hidden dynamics, interdependencies, and feedback loops can either support or stall growth, and how understanding these factors can give you the clarity to lead with greater confidence. Our mission is to help you recognize and navigate the dynamic interactions and hidden structures that influence your success.

Alice Has an Idea and a Company Is Born

Alice had always been wired for entrepreneurship. Even early on, she had a knack for spotting opportunities and designing clever solutions. In grade school, she'd turned her love for Pokémon into a thriving venture, trading strategically to build an enviable collection. By high school, she was running a small but profitable Mother's Day flower business, orchestrating a network of friends and suppliers to help her classmates' dads make the perfect impression. Alice didn't just seize opportunities; she built simple solutions that made things work, often in places others hadn't even noticed were broken.

Now, years later, Alice found herself sketching out a new idea—this time, a solution to a problem she felt every day. She struggled to keep track of birthdays, special dates, and the small but important details that keep relationships strong. What she envisioned wasn't just another app but a connected system designed to manage the complex web of personal connections in her life. It would gather information, surface reminders at just the right time, and help her stay meaningfully engaged. To Alice, it was more than a product; it was a way to bring people closer, one thoughtful moment at a time, while the complexity of our lives swirled around us.

With savings in the bank earmarked for building a minimum viable product (MVP), Alice was ready to bring her idea to life. However, she'd never created a software product before and thought it wise to seek advice. She reached out to her uncle, an experienced software engineer, to ask for guidance on what she should do next.

Her uncle, recognizing the potential in her idea and wanting to ensure she got the best start possible, referred her to his friend Theo, the best software engineer he knew. Theo had a reputation for being not only highly skilled but also incredibly passionate about innovative projects.

Alice contacted Theo, explaining her vision for the product. To her delight, Theo was excited about the idea and agreed to work on it part-time, with the hope of eventually joining full-time if things went well. This collaboration turned out to be one of the best decisions Alice made for her business. Theo's expertise and enthusiasm helped them make significant progress in the first four months. They achieved their MVP quickly, and Alice spent only 75% of the money she'd planned to invest in this phase.

Alice felt on top of the world. She had a great product and an exceptional developer by her side. Everything seemed to be falling into place perfectly. With her MVP ready and resources to spare, she wondered, What could possibly go wrong?

Alice's excitement grew as she prepared to launch her product. She had a few friends and family beta test the software, and everyone found it incredibly useful, integrating it into their daily routines. Based on this positive feedback, Alice felt confident about launching the product to the public.

With the money saved from the initial software development phase, Alice was able to allocate more funds to advertising and marketing efforts. This investment paid off when, within two months, she signed up 50 new users.

The launch wasn't perfect (there were some hiccups and minor issues), but Theo was right there by her side. He worked tirelessly to fix problems and ensure the customers had a smooth experience. His dedication and skills shone through, proving him to be a stand-up guy.

Alice couldn't help but feel incredibly fortunate to have Theo as her partner in this venture. Together, they combined visionary ideas with technical expertise, creating a partnership that felt dynamic and full of potential. Their collaboration was off to an exciting start, and Alice felt a surge of optimism about the future.

After a few months of steady growth, Alice started receiving requests from customers for improvements to her software. Thanks to Theo's skills, they could be very responsive to feature requests, and the customer base continued to grow. Alice couldn't believe her luck. The codebase was solid, bug fixes were quick, and customers were happy. It seemed like everything was going perfectly.

However, as every seasoned entrepreneur understands, the road ahead would bring its share of unforeseen challenges and opportunities. They were only just beginning to explore what lay ahead. A few months into this growth phase, Alice began to notice a troubling trend. Theo, who had been a whirlwind of productivity, seemed to be slowing down. He was still the same enthusiastic and talented software engineer, just as passionate about the product as ever, but features were launching more slowly. He had the energy, the mojo, the skills... but the code wasn't making it to the customers as fast as it used to. The backlog of tasks was building up.

Concerned, Alice approached Theo. "What's going on, partner? We used to ship all these features, and now the backlog is building up."

Theo sighed, looking a bit weary. "Alice, there's just a lot of work to do. More users mean more features, leading to more bugs to find and fix.

With each new feature, it takes longer and longer to test the code, and then with each release, there's more code to push out. It's not that I'm not trying, but the workload has increased significantly."

What's Really Going On Here?

Alice was beginning to sense that the path forward might not be as straightforward as it had first seemed. The initial excitement of growth was giving way to new challenges—more users, more feature requests, and a few bugs starting to crop up. The once-smooth rhythm of progress was beginning to feel a bit more complicated.

What Alice didn't yet realize was that these challenges were just the surface. Beneath them lay deeper forces shaping the destiny of her company. To thrive in this next phase, she would need to uncover and navigate these hidden dynamics.

Unknown to Alice, a system was quietly shaping every step of her company's journey. It was there, whether she saw it or not, influencing outcomes in ways she had yet to understand.

Liquid

PART 2:

There Is Always a System

I know exactly what you mean. Let me tell you why you're here. You're here because you know something. What you know you can't explain, but you feel it. You've felt it your entire life, that there's something wrong with the world. You don't know what it is, but it's there, like a splinter in your mind, driving you mad. It is this feeling that has brought you to me. Do you know what I'm talking about?

–Morpheus, *The Matrix*

Liquid

Chapter 6:
Complexity in Everyday Life

At first, life seems simple. You make a plan, execute it, and expect results to follow. But as soon as you try to coordinate with others, adapt to changing circumstances, or manage competing priorities, that simplicity quickly unravels. Addressing complexity isn't just about making more things happen. It's about understanding the way those things interact, often in unpredictable and surprising ways.

Imagine organizing a family dinner. At its core, it seems straightforward: decide on a date, invite everyone, and cook a meal. But soon, schedules collide. One relative can only come on a Saturday, while another insists Sundays are better. Someone is vegetarian, and someone else has a peanut allergy. You need to coordinate the menu, transportation, and timing, all while dealing with last-minute cancellations or unexpected guests. Suddenly, what seemed simple becomes a tangle of decisions, trade-offs, and compromises.

Complexity doesn't simply add more work; it changes the work itself. It introduces dependencies that we have to manage. If the main dish isn't ready on time, the entire dinner is delayed. If one guest can't make it, it might affect the seating arrangements, the conversations, or even the mood of the evening. The pieces don't exist in isolation; they interact in ways that create ripple effects, both seen and unseen. This same principle applies to our professional lives.

What starts as a straightforward task can quickly escalate when new variables are introduced. A small product update might seem like a quick win, but it requires coordination across design, engineering, and marketing teams. It depends on approvals, testing, and deployment schedules. One unexpected delay in testing could disrupt the entire release timeline. Even with the best planning, surprises emerge because no plan can account for every interaction.

Complexity is a natural part of life. As more people, processes, and moving parts come into play, predicting outcomes becomes increasingly difficult. But we shouldn't see complexity as inherently bad; it's the price of growth. It emerges when we take on meaningful challenges that exceed the capacity of any one person or team to manage alone.

Complexity isn't a failure; it's a sign of progress. The challenge isn't to eliminate it but rather to navigate it effectively. Success comes from recognizing complexity as an integral part of evolution and developing the skills to work within it. In the chapters ahead, we'll explore how this unfolds in Alice's journey and how the hidden systems shaping her company mirror the ones at work in our own lives.

The Rise of Complexity in Growing Companies

What starts as a clear and focused effort with a small team working closely together can quickly expand into a maze of interconnected challenges.

In organizations, complexity emerges as teams grow, products expand, and customer demands increase. This is a natural side effect of growth. Imagine Alice's growing company. At first, everything felt manageable. Her small team could make decisions quickly, resolve issues in real time, and stay aligned on priorities. But as new hires joined, new processes were introduced, and the product evolved, things began to shift.

A bug fix in one part of the product started causing unexpected side effects elsewhere. A new feature meant marketing needed materials, sales needed training, and support needed documentation, all while the engineering team was working to hit a tight deadline.

This kind of complexity is normal in any growing business. Each new addition, whether a person, a feature, or a customer, creates new interactions and dependencies as well as adding more to manage. These layers of complexity often emerge unnoticed at first, but over time, they shape the rhythm of the entire organization.

Understanding this transition is the first step in learning how to navigate it. Complexity in organizations isn't just about adding more simplistic tasks to the mix. Complexity arises as a result of how everything connects and the unexpected ways in which those connections can influence outcomes.

Liquid

Chapter 7:
Understanding Complexity

As Alice and Theo begin to feel the weight of growing complexity, they're not alone. Every business, regardless of size, reaches a point where simple solutions no longer suffice and unseen forces shape outcomes in unexpected ways. The challenge isn't just dealing with complexity—it's recognizing it in the first place. Before we can manage complexity effectively, we must learn to see it, understand its patterns, and uncover the hidden dynamics that influence every decision and outcome.

As businesses grow, complexity is inevitable. It's not a failure of leadership or process but a natural outcome of success. Every achievement brings new connections, dependencies, and interactions, weaving the organization into an increasingly intricate system. Like threads in a spider's web, each new hire, product, market expansion, or process change strengthens the business—but it also creates unseen points of interaction that can ripple across teams in unpredictable ways.

At first, complexity emerges subtly. A team adds a new workflow to streamline decision-making, only to find that it slows collaboration elsewhere. A product update designed to improve user experience unexpectedly disrupts customer support operations. Leaders implement a reporting structure to ensure alignment, only to realize that the layers of approvals now hinder agility.

These shifts don't appear as single, isolated problems but as symptoms of a deeper, interconnected system at work. The challenge isn't just that complexity exists—it's that it often operates invisibly until it manifests as bottlenecks, misalignment, or inefficiencies.

Without recognizing its patterns, leaders may find themselves solving surface-level issues while deeper systemic challenges remain unresolved. Before complexity can be managed, it must first be seen. Understanding how these connections evolve, where dependencies form, and how small decisions influence the larger system is the first step toward navigating complexity effectively.

The pioneering work of Donella Meadows in *Thinking in Systems*[8] highlights that complexity isn't about chaos. Instead, it's about the relationships between the elements of a system. Meadows emphasizes that systems behave in patterns, and while they may appear unpredictable, their underlying dynamics often follow certain consistent principles. This understanding is crucial for leaders who want to navigate complexity effectively.

Why Complexity Matters

In growing businesses, complexity doesn't simply add more tasks. Complexity changes how decisions must be made. Leaders who focus only on surface-level issues often find themselves in a reactive cycle, constantly firefighting symptoms rather than addressing the root causes.

Peter Senge, in *The Fifth Discipline*,[9] emphasizes that real progress comes from seeing the "whole system" and identifying leverage points where small adjustments can have outsized impacts.

For example:

- **Technical complexity:** An engineering team deploys a new feature, only to discover it slows down other parts of the system due to hidden interdependencies. The team scrambles to patch the issue while users grow frustrated.

- **Organizational complexity:** As teams grow, communication silos form. One group works on a feature already being developed elsewhere, leading to duplication and wasted effort.

- **Market complexity:** A competitor's unexpected move forces a sudden shift in strategy, disrupting existing plans and creating confusion across departments.

These scenarios involve solving immediate problems but also understanding the system-wide effects they represent. In complexity, a change in one part of the system can ripple across the entire organization.

Complex Adaptive Systems in Action

Organizations are not static. Organizations are living, breathing systems that adapt to their environment. Known as **complex adaptive systems (CAS)**, these entities are shaped by countless interactions, where the whole is greater than the sum of its parts.

As John Holland explains in *Emergence: From Chaos to Order*,[10] CAS are characterized by their ability to evolve—not through rigid top-down control but through continuous adaptation based on feedback.

Feedback, in this context, goes beyond data or direct input. It's the accumulation of signals from customers, employees, market shifts, and internal operations that shape decision-making. In a well-functioning organization, feedback loops guide adaptation, helping teams refine processes, improve products, and align strategies with evolving conditions.

For instance, an engineering team deploying a new feature may receive user feedback indicating unforeseen friction, prompting iterative changes to enhance usability. A sales team struggling with customer objections might adjust their messaging based on recurring patterns in prospect conversations. Even internal structures evolve in response to friction—when leaders recognize bottlenecks in decision-making, they might refine reporting lines or empower cross-functional teams to work more autonomously.

The challenge isn't just receiving feedback. It's recognizing which signals matter, interpreting them correctly, and acting on them effectively. Strong organizations cultivate mechanisms to detect, analyze, and incorporate feedback at every level, ensuring that adaptation isn't reactive chaos but an intentional process of refinement and growth. In a CAS, change is constant, and those who learn to harness feedback as a guiding force will navigate complexity with greater agility and resilience.

Alice's growing company is a perfect example. At first, her small team moved quickly and stayed aligned. But as new hires joined and the product expanded, unanticipated challenges emerged. Dependencies between teams created bottlenecks, small bugs escalated into major problems, and priorities became harder to manage. What Alice didn't realize was that these struggles weren't isolated. They were the visible effects of a deeper system quietly shaping her company's trajectory as it grew.

The Path Forward

Mastering complexity isn't about reducing it but about learning how to move through it with clarity and intention. Leaders who develop the ability to see the hidden systems shaping their organizations can step beyond reactive problem-solving and into proactive influence. Instead of treating complexity as a barrier, they can use it as a strategic advantage—understanding its patterns, leveraging its feedback loops, and designing systems that evolve rather than break under pressure.

As Donella Meadows points out, "The behavior of a system cannot be known just by knowing the elements of which the system is made." True understanding comes from seeing how those elements interact, adapt, and generate outcomes that no single component could produce alone. This shift in perspective—from managing isolated parts to shaping interconnected forces—creates a path toward more sustainable decision-making, where leaders guide the system rather than merely reacting to it.

Looking ahead, the most effective leaders will be those who embrace systems thinking as a core competency. They will recognize that every decision, no matter how small, sends ripples through their organization, influencing both immediate results and long-term adaptability. By fostering resilience, designing for flexibility, and understanding the forces at play beneath the surface, they can prepare their businesses not just to survive complexity but to thrive within it. The future belongs to those who see complexity not as a burden but as a tool for shaping what comes next.

Chapter 8:
Recognizing Patterns of Complexity

A complex adaptive system (CAS) experiences many types of complexity. Companies are CAS. Within companies, we most often see the following types of complexity, which we'll define in detail below:

- technical complexity

- organizational complexity

- domain complexity

- market complexity

- customer complexity

- regulatory complexity

Naming the types of complexity we experience within our businesses is a powerful first step toward understanding and managing them. When things feel nebulous and overwhelming inside a company, it can paralyze decision-making and lead to reactive, short-term fixes.

However, when we can identify this as a symptom of complexity and give that complexity a name (e.g., technical, organizational, etc.), we make it tangible. With a name, complexity becomes something we can observe, discuss, and, ultimately, influence. It's no longer an abstract problem. Instead, it's a specific challenge we can see.

When we can see the pattern, we can take steps to address it. By identifying the type of complexity at play, we can apply targeted patterns or playbooks to influence the system to move toward a more balanced state. For example, technical complexity stems from a growing codebase and might be addressed with a refactoring playbook, introducing modularity to reduce dependencies and simplify future development. Organizational complexity that's driving communication overload might be treated with a pattern that defines clearer boundaries for collaboration and prioritizes signal clarity over noise.

Recognizing the types of complexity also helps us avoid one-size-fits-all solutions that fail to address root causes. Each type of complexity has unique characteristics and requires tailored interventions. The same principle applies across all aspects of a business, from engineering to sales to operations.

By categorizing complexity and matching it with proven patterns, leaders can take deliberate, informed actions to influence the system, moving it toward a flow state where the organization operates with clarity, adaptability, and alignment. Through naming and understanding, we transform complexity from an obstacle into an opportunity for growth and innovation.

Understanding Complexity Types

Understanding these patterns is the first step toward managing the hidden systems that shape every decision and outcome.

Within companies, we most often see the following types of complexity.

Technical complexity arises from the interconnected nature of systems, processes, and infrastructure. As a company grows, even small changes in one area can have unintended consequences across the organization.

Example: Alice's engineering team deploys a new feature to the software, but the update unexpectedly slows down critical functionality due to hidden dependencies. Debugging and resolving the issue requires cross-team collaboration and a deeper understanding of how different systems interact.

Managing technical complexity requires robust processes for testing, documentation, and anticipating the cascading effects of change.

Organizational complexity develops as teams expand, roles diversify, and decision-making pathways multiply. What works for a small team often becomes cumbersome in a larger organization, leading to inefficiencies and misalignment. The complexity that arises within company culture is a form of organizational complexity.

Example: Alice notices that her growing product and sales teams aren't coordinating effectively. Sales promises features to clients that are still in development, creating frustration and rushed timelines for the product team.

Addressing organizational complexity involves establishing clear roles, creating transparent communication channels, and ensuring alignment across all levels of the organization.

Domain complexity reflects the intricacy of the industry or sector in which a company operates. It arises from the specialized knowledge required to solve problems or create value within a specific field.

Example: Alice's company operates in healthcare software, requiring her team to understand complex medical terminology and workflows. A misunderstanding of key domain requirements leads to a feature that fails to address real user needs.

Navigating domain complexity demands deep expertise, ongoing learning, and collaboration with subject-matter experts.

Market complexity stems from external factors such as competition, economic conditions, and shifting customer expectations. Companies must continuously adapt to stay relevant in unpredictable environments.

Example: A competitor releases a similar product feature, undercutting Alice's pricing strategy. Her team scrambles to differentiate their offering and communicate its value to customers.

Adapting to market complexity requires agility, a focus on innovation, and the ability to anticipate and respond to external trends.

Customer complexity arises from the diverse needs, behaviors, and expectations of a company's user base. It reflects the challenge of satisfying a broad spectrum of requirements while maintaining a coherent product strategy.

Example: Alice's software is used by small businesses and enterprises, each with vastly different demands. Catering to one group risks alienating the other, leaving Alice's team torn between competing priorities.

Managing customer complexity involves understanding user personas, prioritizing features effectively, and delivering value that resonates with diverse audiences.

Regulatory complexity refers to the need to comply with laws, standards, and industry-specific regulations. As businesses scale, the burden of compliance can grow exponentially.

Example: Alice's team discovers that a planned feature violates data privacy laws in a key market, requiring them to redesign the feature and delay its release.

Navigating regulatory complexity requires staying informed about changing rules, building compliance into processes, and investing in expertise to mitigate risks.

Seeing the Bigger Picture

Each type of complexity exists within a broader system and rarely acts independently. Technical challenges can amplify organizational inefficiencies, market shifts can create customer demands that seem impossible to balance, and regulatory changes can disrupt carefully planned strategies.

Together, these patterns of complexity create the web of interconnections that leaders must navigate. Recognizing and addressing these patterns is essential for sustainable growth. By identifying the root causes of complexity and understanding how different types interact, leaders like Alice can move from reactive problem-solving to proactive systems thinking, paving the way for long-term success. Let's dive into some deeper examples to build our skills in identifying the patterns of complexity at play and how they interrelate.

Interdepartmental Collaboration

In a company, different departments, such as marketing, sales, product development, and customer support, are all *elements* within the larger system that is your business. Each department has its own objectives, processes, and metrics for success, but they're interrelated and interact constantly.

For instance, a new product feature is defined by product and developed by the engineering team; it will require marketing to create promotional materials to advertise it, sales to adjust their demo strategies to incorporate it, and customer support to handle increased inquiries after customers start using it. This is *organizational complexity*. The way in which these departments interact can lead to *emergent behaviors*. Emergent behaviors are unpredictable outcomes that arise from the interactions of multiple elements within a system rather than from any single decision or action.

For example, a shift in marketing strategy might unexpectedly drive demand from an unintended customer segment, or subtle changes in product features could trigger new customer usage patterns that weren't anticipated. These emergent behaviors, in turn, increase *customer complexity*—the evolving, often unpredictable nature of customer needs, behaviors, and expectations. As companies grow, their customer base diversifies, making it increasingly difficult to anticipate how different segments will respond to changes in products, messaging, or market conditions. Recognizing and adapting to these emergent behaviors is essential for navigating customer complexity effectively.

These outcomes are often not predictable from the actions of any single department alone. Likewise, when things go awry, it's rarely due to activity within just one department. These departments need to operate together, in harmony, over time, to achieve the desired outcomes.

Market Dynamics and Customer Behavior

A company operates within a market that in and of itself is a complex system. The market consists of various players, competitors, customers, and suppliers, each making decisions that affect others.

For example, a price change by a competitor could lead to shifts in customer preferences, which then affect the sales and marketing strategies of your company. Customer behavior, influenced by marketing campaigns, word-of-mouth, and market trends, often results in emergent patterns, such as viral product popularity or sudden drops in demand. These patterns are complex and cannot be easily predicted by looking at individual customer actions in isolation. This is *market complexity*.

Company Culture

The culture of a company is another example of *organizational complexity*. It's shaped by the interactions among employees, the leadership style, company policies, and external influences.

A culture of innovation might emerge not from a top-down directive but from the way in which teams collaborate, share ideas, and respond to successes and failures. This culture, in turn, influences employee behavior in ways that aren't easily predictable from the individual elements alone, such as the hiring practices, the design of office spaces, or the company's mission statement.

The complexity of company culture is why it's so difficult to shift it. We cannot just hold a meeting to tell people to behave differently. The nebulous "culture" that we try to steer is an emergent behavior produced by long-term, continuous interactions across the company. Shifting a culture requires us to work to *influence* the culture to evolve; we can do this by putting in place boundaries and aligning interactions on a consistent, long-term basis through clear processes and accountabilities.

This is why we often say that the culture of a company starts at the top with the CEO. The boundaries and interactions that the CEO chooses to implement to influence the system will determine what sort of company culture emerges. The CEO sets the tone not just through formal policies or direct communications but also through their everyday actions, their priorities, and the signals they amplify or suppress.

These decisions and behaviors create a framework within which teams operate, shaping how employees interact, solve problems, and align with the company's goals.

Unfortunately, when the CEO is loose or inconsistent in these areas, failing to establish clear boundaries or sending conflicting signals, the organization often experiences chaos or stagnation. A lack of clarity at the top leads to teams working at cross-purposes, confusion over priorities, and a fragmented culture that erodes trust and alignment.

For example, a CEO who frequently changes direction without clearly communicating why creates an environment where employees are unsure of their focus or role in achieving the company's mission. This uncertainty breeds inefficiency and frustration, preventing the system from operating in a state of productive flow.

Culture doesn't develop in a vacuum. When the CEO is inattentive to the boundaries and interactions within the company, the resulting culture will often default to the path of least resistance, which can mean chaos, rigidity, or a patchwork of misaligned behaviors across teams. But just as culture can drift into dysfunction without guidance, it can also be intentionally shaped with the right leadership.

When a CEO actively and thoughtfully influences the system, they create an environment where employees feel empowered, focused, and aligned. The culture becomes an intentional force for cohesion and growth rather than an unintended byproduct of neglect or inconsistency. In this way, the CEO's influence isn't just foundational; it's transformative.

Coadaptation in Product Development

Innovation within a company isn't a linear process. It emerges from the dynamic interactions between various elements, such as employee creativity, research and development processes, cultural norms, market feedback, and resource availability. These elements don't operate in isolation. Instead, they interact in ways that create ripple effects throughout the organization, shaping how new ideas evolve into viable products or services.

For innovation to thrive, the organization must be sufficiently connected, fostering collaboration and trust. This environment allows ideas to cross departmental boundaries, where an initial concept in one team sparks a related insight in another.

Through discussion, refinement, and shared problem-solving, what starts as a small idea can gain traction, turn into a well-structured plan, and ultimately lead to new product development. This interplay is an example of **coadaptation**, a process in which different aspects of an organization evolve together in response to changes in one another. Coadaptation ensures that as one part of the system changes, such as the introduction of a new product feature, the surrounding elements adjust in a way that maintains balance and functionality. Without this mutual adjustment, growth in one area can create stress on the overall system, leading to inefficiencies or missed opportunities.

For example, when a company develops a new product feature based on customer demand, this change drives *technical complexity* as engineers implement new systems to support it. In turn, technical advancements often necessitate *organizational complexity*, requiring new release management workflows, new roles, or even structural adjustments to manage the added demands. If these organizational changes don't keep pace with technical growth, the company risks being underresourced or unable to sustain its technology effectively.

Beyond internal adjustments, *customer complexity* also co-evolves with technical advancements. A new feature designed to meet an immediate customer need might lead to unforeseen secondary requests or shifts in customer behavior, further influencing product direction. This continuous cycle of feedback, adaptation, and refinement highlights the interconnected nature of innovation within a complex system.

Recognizing and fostering coadaptation is essential for organizations aiming to sustain innovation while managing complexity. By acknowledging that changes in one domain inevitably impact others, companies can proactively design systems that accommodate mutual adaptation.

This approach prevents bottlenecks and misalignment, ensuring that growth remains sustainable across all facets of the organization. Ultimately, innovation in a complex system is an emergent property— one that results from the interactions between multiple, evolving elements rather than a singular directive.

Companies that embrace this perspective position themselves to not only create new products but to evolve in tandem with their own expanding complexity, ensuring long-term adaptability and success.

Crisis Management and Response

When a company faces a crisis, such as a data breach or a public relations issue, the response involves multiple departments and rapid interactions between them. The way in which the crisis is handled can lead to emergent outcomes such as changes in customer trust, regulatory scrutiny, or shifts in company policy that weren't predictable at the onset of the crisis.

The complexity of these interactions and their outcomes highlights how this complex system operates within a company, where the collective behavior during a crisis isn't just the sum of individual actions but something new that emerges from the system's response (or lack thereof). This example also underscores the importance of *coadaptation* within a company. As the crisis unfolds, *customer complexity* (shifts in trust), *organizational complexity* (policy changes or team realignments), and *regulatory complexity* (new scrutiny or compliance requirements) evolve together, each influencing the others. The company's ability to adapt effectively depends on how well these types of complexity align and respond to one another.

What's Going On for Alice and Theo?

Adapting to complexity is an ongoing challenge, one that Alice and Theo are about to experience firsthand. Alice simply thinks that Theo is "coding more slowly"; however, an entire ecosystem of complexity is beginning to take form beneath the surface. Because of this, Theo is now struggling to predict how long a feature will take to complete, and there is no one simple change he can make that will bring him back to a state where he can easily know the answer.

Theo isn't just coding more slowly. He's grappling with a growing tangle of *technical complexity* that's slowing him down in ways he didn't expect. At first, adding features felt straightforward. But as the codebase expands, each new addition interacts with an increasing number of existing components, making changes harder to predict and test. Debugging simple issues now requires navigating layers of dependencies. The once-quick process of deploying updates has become more fragile, requiring extra checks to ensure that one fix doesn't break something else. Theo finds himself spending more time reading old code than writing new code, manually testing changes that used to feel effortless, and juggling competing priorities as unexpected bugs emerge from areas he hadn't even touched. What used to be smooth, rapid progress has turned into a slow, careful dance with an increasingly complex system.

The more components and interactions there are within a system, the more complex the system becomes. Let's take a look at where Alice and Theo's software application currently stands.

At first glance, it might seem simple in that it just helps users manage their contacts and reminders. But behind the scenes, there's a lot happening. The app must communicate with servers to sync data, manage and store user information securely, and ensure that new features don't interfere with existing ones. Each of these tasks involves layers of code, hardware, people, and processes working together.

Is Theo truly just "coding more slowly," or is he uncovering a deeper challenge within their growing system? What will it take to regain momentum and overcome the complexity holding him back?

Let's find out.

Chapter 9:

Alice Doubts Theo

Concerned, Alice approaches Theo. "We used to ship all these features, and now the backlog is building up. What do you think is causing this?"

Theo sighs, looking a bit weary. "Alice, there's just a lot of work to do—more users mean more features, more bugs to find and fix."

Alice furrows her brow. "Are you feeling overwhelmed, or is the product getting too complicated for your skills?"

Theo shakes his head. "It's not that. It's just the nature of the work. As we add more features, the codebase is growing, and each change gets more difficult to manage. Additionally, each new feature has to be tested against everything else to ensure it doesn't break something somewhere else."

Alice is getting impatient. She doesn't want excuses; she wants speed. "I just want to understand why things are slowing down. I know you're capable, Theo, but it feels like something's changed for you. Are you just not interested in our project anymore?"

Theo pushes back: "I get it, Alice. It's not about losing interest or slacking off. The manual testing process is time-consuming, and as the codebase grows, it's taking longer to ensure everything works perfectly. Plus, deploying the code manually adds more time and potential for errors."

Both are feeling exasperated. Neither one of them feels heard. They agree to take a breather and pick the conversation up at their next 1:1.

Alice and Theo are facing a significant challenge, and they're about to fall off a cliff.

Theo's Perception

Theo's hypothesis is that he simply has more work as the system grows. He notices himself spending more time at the keyboard and assumes the workload has increased. While the workload has increased, this isn't necessarily the complete truth as to why he's feeling the strain.

Theo's situation mirrors Morpheus in *The Matrix*.[11] Morpheus understands the hidden forces at play but struggles to explain them to someone who's only ever known the surface reality. Morpheus can't simply tell Neo what the Matrix is; Neo has to experience it for himself.

Likewise, Theo senses that coding is taking longer, simple tasks feel more complicated, and unexpected issues keep surfacing, but he can't clearly articulate why. He knows he's working just as hard, yet his explanations to Alice come across as vague and defensive because he lacks the language to describe the invisible complexity slowing him down. Like Neo, Theo must learn to see the deeper system at play before he can begin to navigate it effectively.

This inability to explain the situation is akin to the complexity of understanding the Matrix itself. The Matrix is an all-encompassing illusion that's difficult to describe because it permeates every aspect of Neo's perceived reality.

Similarly, the growing complexity of Theo and Alice's software project is pervasive, affecting everything Theo does, yet a clear, single root cause remains intangible and elusive.

Later in this story, we'll help Theo realize what's actually going on. But for now, let's look at what Alice is thinking.

Alice's Perception

None of what Theo is experiencing within this complex system is even within Alice's line of sight. To Alice, the CEO, these underlying challenges don't exist because they're hidden within the system boundaries of the technology, which is an area she cannot directly observe or interact with.

Liquid

Chapter 10:
The Silent Cost of Growing Complexity

Humans naturally struggle to understand things they can't see or directly measure. We rely heavily on tangible signals like data, reports, deadlines, and outcomes to form our understanding of how things work. When the complexity of a system is invisible, like the intricate dependencies and interactions within Theo's technology domain, it's easy to assume everything is functioning as expected until something breaks.

Why Complexity Grows

As systems evolve, new features and functionalities are added to address user needs and stay competitive. Each new feature introduces additional components and dependencies, creating a more intricate web of interactions.

For example, when Alice's software product adds features like event reminders or integrations with social media platforms, these additions require new code modules, data flows, and interactions. Each of these must seamlessly integrate with the existing system, creating new layers of complexity.

This process is often accelerated by the pressure to deliver quickly. In the rush to meet early market demands, teams may have to implement temporary solutions or take shortcuts to ship features on time. These choices, while practical in the moment, *accumulate* over time and form what's known as technical debt. This debt doesn't just disappear; it becomes a growing burden that each successive development effort must navigate or address.

This struggle stems from our inherent need to simplify the world around us. Humans are wired to seek patterns and clarity, but a complex system often operates in ways that defy straightforward logic. In *Thinking, Fast and Slow*,[12] Daniel Kahneman explores how our cognitive biases lead us to favor simple explanations over complex realities, often oversimplifying intricate systems.

John Gall's *Systemantics*[13] delves into the pitfalls of oversimplification in system design, highlighting the unintended consequences that arise when we underestimate complexity. When leaders view a system as simpler than it truly is, they risk implementing solutions that fail to address root causes, inadvertently creating new problems instead of solving existing ones.

Oversimplification is particularly detrimental in business because it leads to decision-making based on an incomplete understanding of how different parts of the system interact. Leaders may implement surface-level fixes without realizing how deeply interconnected their operations are. For example, mandating a strict productivity metric, like measuring engineering success purely by lines of code written, might seem like a way to drive efficiency; however, it incentivizes developers to prioritize quantity over quality, leading to bloated, unmaintainable software that slows progress in the long run.

When leaders fail to account for system complexity, they create policies and structures that seem logical in isolation but generate hidden friction, inefficiencies, or even outright failures over time. Recognizing and respecting complexity allows businesses to craft solutions that align with how systems truly behave rather than how we wish they would.

Complex systems are challenging because of their nonlinear nature. Small changes can trigger disproportionately large effects, and actions that seem straightforward often have unintended consequences.

For example, a company might decide to speed up development by cutting back on code documentation, believing that experienced engineers will fill in the gaps. In the short term, this saves time, but as the team grows, new engineers struggle to understand legacy systems, slowing down future development and increasing onboarding time.

Other shortcuts taken during development may seem harmless in the moment but can amplify over time, eventually creating highly manual and inefficient processes for the customer service team to manage downstream.

Similarly, an organization might rush a new feature to market without fully testing its integration with existing workflows, only to find that it disrupts customer usage patterns, overwhelms support teams with complaints, and requires costly fixes. Leaders who fail to account for this nonlinearity may find themselves puzzled by the unpredictable outcomes of their decisions. Understanding this principle allows leaders to approach complexity with humility, resisting the temptation to succumb to the allure of simplicity.

True leadership lies not in trying to control every outcome but in influencing the system as it evolves. This shift in perspective empowers us to guide the system more effectively, aligning it with our short- and long-term goals.

Alice might look at metrics like delivery timelines or defect rates, but those numbers don't reveal the deeper system dynamics such as cascading dependencies in the codebase or the increasing cognitive load on Theo's team. Without visibility into these layers, Alice's perspective on the situation is incomplete, leaving her to unknowingly address symptoms rather than root causes. Or, worse, Alice will make judgments and decisions based on faulty or incomplete data.

Alice's simplistic view of the complex system of product engineering is that she injects ideas into the process and features come out of it. Because the technical complexity lives within the boundaries of the complex system, she cannot see it from the outside. She only sees the negative emergent behavior that is caused by that complexity, which is that feature release is slowing down. And then she makes an assumption based on the symptom that she can see.

This disconnect highlights a critical challenge in leadership: the ability to effectively lead systems we cannot fully see or directly control. It requires trust, collaboration, and, most importantly, the willingness to listen to those who are closer to the system's boundaries. For Alice, recognizing this gap in her understanding is the first step toward influencing the system effectively.

By seeking to understand the unseen forces at play and collaborating with Theo to uncover the system's dynamics, Alice can begin to align her actions with the realities of the system rather than the surface-level indicators she's been relying on. Alice and Theo are in a state of confusion, much like Neo before he fully understands the Matrix. They sense that something is wrong; the system doesn't fit their expectations. They can feel the disconnect, but they can't quite clearly articulate what the issue is.

This sense of unease and the suspicion that there's more beneath the surface requires curiosity, patience, and an open mind for both Alice and Theo to uncover the truth.

If we could parachute into their world and talk to Alice and Theo at this point in their experience, we would sound like Morpheus in *The Matrix*, revealing the hidden complexities and unseen forces at play. However, because they can't yet see these things, they would be resistant to believing us.

The Accumulation of Complexity Over Time

The challenges Alice and Theo face aren't unique to their team. Complexity in a business doesn't dissipate when a task is completed or a feature is delivered. Complexity accumulates. This fundamental misunderstanding often creates tension between CEOs and CTOs, as they view the nature of complexity through entirely different lenses.

From the CEO's perspective, complexity might seem like a temporary challenge. Once the engineering team completes a feature, the CEO may assume that the associated complexity has been "managed" and that the business can now move forward unburdened.

After all, the feature is live, customers are using it, and the team can focus on the next task. The CTO, however, sees a very different reality. While the initial complexity of building the feature may have been addressed during the development phase, the real complexity often emerges after delivery.

The new feature doesn't exist in isolation; it interacts with the broader system. These interactions introduce unforeseen interdependencies that weren't fully apparent during development. For example, a feature designed to improve the customer experience might inadvertently introduce latency in the system or create challenges for future integrations.

The CEO's perspective tends to focus on the visible outcome—the completed feature—while the CTO grapples with the hidden complexity that's now become part of the system. To the CTO, the work isn't finished once the feature is delivered; it's just begun. The feature must now be maintained, integrated with other systems, and continually monitored to ensure it doesn't create cascading issues.

This disconnect highlights a critical leadership challenge: aligning the CEO's strategic goals with the CTO's understanding of the system's long-term health. A key part of this alignment involves educating stakeholders on the nature of accumulating complexity and the importance of addressing technical debt proactively. Without this shared understanding, the business risks reaching a tipping point where accumulated complexity slows progress to a crawl, stifling innovation and growth.

To manage this accumulation effectively, CTOs must advocate for practices like refactoring, automated testing, and regular system reviews, which are discussed more deeply later.

These efforts, while not always visible to the CEO, are essential investments in the company's ability to adapt and thrive over time. By bridging the gap in perspective, leaders can ensure that complexity is not only managed but also anticipated and minimized as the company grows.

Now, let's get back to Alice and Theo's dilemma.

Liquid

Chapter 11:
Understanding Phase Changes

The increase in complexity and the divergence of perspectives signifies that Alice and Theo are experiencing a phase change in their company, yet they haven't cognitively realized it has happened.

These phase changes can be likened to the familiar phases of water. Just as water doesn't instantly change from room temperature to boiling or freezing, their project hasn't suddenly been overwhelmed by complexity. The complexity has always been gradually increasing; they just haven't recognized it yet. The complexity of their project will gradually reach critical points where the dynamics begin to shift, eventually leading to chaos or stagnation if not properly managed.

A complex system is ordinarily in one of three states:

- **Frozen (solid):** A rigid state where the business is overregulated and resistant to change, leading to stagnation and inefficiency.

- **Liquid (flowing):** A balanced state where the business operates with agility and adaptability, efficiently responding to changes and fostering continuous growth.

- **Boiling (steam):** A chaotic state where the business is overwhelmed by complexity, with uncontrolled, erratic behaviors that disrupt operations and stability.

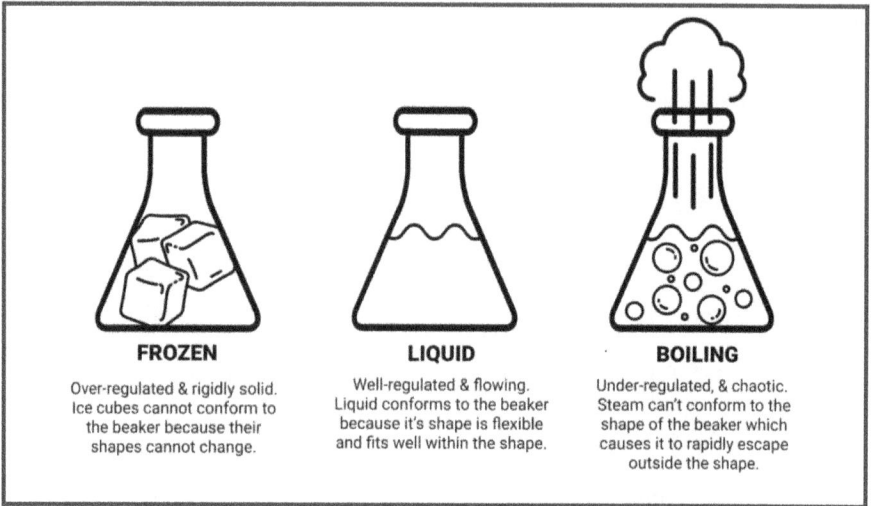

FROZEN	LIQUID	BOILING
Over-regulated & rigidly solid. Ice cubes cannot conform to the beaker because their shapes cannot change.	Well-regulated & flowing. Liquid conforms to the beaker because it's shape is flexible and fits well within the shape.	Under-regulated, & chaotic. Steam can't conform to the shape of the beaker which causes it to rapidly escape outside the shape.

Adaptability

The phase in which a complex system exists is one of the primary factors that determines if the system can adapt to new environments and circumstances. Just like water, a complex system can exist in different states: boiling with chaos, frozen in rigidity, or flowing in a liquid state.

Each phase shapes how the system responds to change, and understanding these phases is crucial for influencing the system effectively:

- A **liquid** state, for example, allows adaptability, where interactions flow smoothly, boundaries are flexible, and the system can adjust dynamically to external pressures or opportunities.

- A **boiling** system is chaotic, with too many interactions happening simultaneously and no clear boundaries to guide behavior. The system is overstimulated and underregulated, leading to inefficiencies and unpredictable outcomes. Boiling systems struggle to adapt because their energy is scattered, with no coherence to direct the system toward meaningful change. This is called decoherence.

- A **frozen** system, while more predictable, is too rigid to respond to new circumstances. Boundaries are overly restrictive, interactions are limited, and the system resists any deviation from established norms.

Boiling destroys the system, and freezing traps the system. Both extremes make adaptation impossible—one through unchecked chaos, the other through rigid stagnation.

The key to adaptability lies in maintaining a liquid state, where structure and flexibility are balanced. In this state, the system can respond effectively to new challenges, shifting without breaking. Interactions are neither chaotic nor overly constrained; instead, they're purposeful and aligned.

But this balance isn't static. It exists along a fine and dynamic threshold known as the *edge of chaos*. In systems theory, the edge of chaos describes the space between complete rigidity and uncontrolled disorder, where complexity is at its most powerful.

This is where businesses naturally operate—between the frozen structures that stifle progress and the boiling chaos that leads to collapse. At this edge, innovation, collaboration, and sustainable progress thrive, but it's also an inherently unstable equilibrium. If left unchecked, a business at the edge of chaos can quickly tip too far in one direction, slipping into freezing stagnation or boiling disorder.

The challenge for leaders is to recognize this fine balance and guide the system to remain in a liquid state. It's at this edge of chaos, where structure and adaptability coexist, that organizations find their greatest potential for resilience, efficiency, and long-term success.

Leaders who understand this principle can focus their efforts on moving their organizations toward a liquid state, where the system's potential for innovation, growth, and resilience is maximized. By identifying whether the system is boiling, freezing, or flowing, leaders can make informed decisions to influence its phase and position their organizations to thrive in dynamic environments.

Let's take a deeper look at what each phase looks like inside a company.

Boiling Point: Transition to Technical Chaos

A "boiling" system refers to the situation where the complexity within an organization has reached a critical point, leading to chaos and inefficiency. This state manifests across various departments and functions, affecting the ability of the company to adapt and grow.

As the organization grows, the number of processes, projects, and interdependencies increases. When the complexity becomes unmanageable, different departments struggle to coordinate effectively. This can lead to duplicated efforts, conflicting priorities, and a lack of clarity about roles and responsibilities. As a result, even simple tasks become difficult to execute, with decision-making processes bogged down by the sheer volume of considerations.

In a boiling system, the likelihood of errors and miscommunications across the company rises significantly. Departments may operate in silos, leading to misunderstandings about goals, strategies, and expectations. This can result in failed projects, missed deadlines, and costly mistakes. A lack of clear communication channels exacerbates the situation, creating a vicious cycle of errors and corrective actions that further complicate operations.

As the organization boils, collaboration deteriorates between departments and teams. The increased complexity makes it difficult for teams to align on priorities, share information, and work together effectively. This can lead to duplicated efforts, conflicting strategies, and a general lack of coherence in how the company operates. As a result, overall productivity declines across the company.

Employees may find themselves spending more time addressing crises and putting out fires than on their core responsibilities. The increased workload and chaotic environment can lead to burnout, lower morale, and higher turnover rates. The sense of constant urgency and pressure undermines the team's ability to focus on strategic initiatives and long-term goals.

In a boiling system, leadership may feel as if they're losing control of the company's operations. As the complexity grows, it becomes harder to predict outcomes and ensure that all parts of the organization are working toward the same objectives.

Leaders may struggle to implement changes effectively as the system's behavior becomes increasingly unpredictable. This loss of control can lead to a reactive rather than proactive management style, where decisions are made in response to crises rather than as part of a strategic plan. The absence of clear communication and collaboration mechanisms further exacerbates the chaos, making it harder to regain control.

A boiling system is often burdened with high levels of organizational debt, which is similar to technical debt in software development. This can take the form of outdated processes, inefficient workflows, and legacy practices that no longer serve the company's needs.

Over time, this organizational debt accumulates, making it harder to implement changes, improve efficiency, or innovate. The company becomes weighed down by its own complexity, struggling to adapt to new challenges and opportunities.

To prevent a company from reaching this boiling point, it's crucial for leaders to recognize the early signs of increasing complexity and take proactive steps to manage it.

This might include adding streamlined processes, clarifying accountability, improving communication and collaboration across departments, and regularly reviewing and updating organizational structures and practices. We'll dive deeper into these suggestions in Part 4 of this book.

Freezing Point: Overregulation and Stagnation

In an effort to control the chaos, some organizations might implement strict protocols that lead to the opposite extreme: overregulation, which can be just as damaging. A "frozen" system refers to a state where the organization has become overly rigid, resistant to change, and stifled by inflexible processes. This rigidity can prevent the company from adapting to new challenges, ultimately leading to stagnation.

A common pattern we see is the CEO becoming too hands-on with the details as a reaction to a boiling state in an attempt to regulate it. In such cases, the CEO's over-involvement in day-to-day tasks and micromanagement can stifle the autonomy of department heads and key decision-makers. This behavior leads to bottlenecks, as every decision must pass through the CEO, slowing down operations and preventing timely actions.

Employees, particularly managers, become disengaged and less likely to take initiative, fearing that their efforts will be overridden or questioned. Over time, this creates an atmosphere of dependency where teams hesitate to innovate or take risks, waiting for the CEO's approval on even minor decisions.

This overcontrol can severely limit a company's ability to adapt, leaving the organization rigid, unresponsive to market changes, and stuck in outdated processes.

In this case, the business quickly moves from boiling to frozen, skipping over the desirable liquid state of *flow*. In a frozen system, the organization is heavily burdened by overregulated processes and excessive bureaucracy. Layers of approvals, rigid workflows, and strict adherence to outdated policies slow down decision-making and stifle innovation.

Employees may find themselves spending more time navigating these procedures than actually performing their core responsibilities. As a result, the company struggles to respond to changes in the market, customer needs, or internal challenges. A defining characteristic of a frozen system is an entrenched resistance to change. Whether due to fear of the unknown, comfort with the status quo, or an aversion to risk, the organization becomes incapable of evolving.

This resistance manifests in a reluctance to adopt new technologies, update processes, or explore new business models. The company becomes stuck in its ways, unable to adapt to shifting market dynamics or emerging opportunities.

In a frozen state, the company experiences a significant decline in innovation and creativity. The rigid structures and fear of change create an environment where new ideas are discouraged and employees are less likely to take initiative. Innovation becomes an afterthought, often blocked by bureaucratic hurdles or dismissed as too risky. This lack of creativity can lead to the company falling behind competitors who are more agile and open to experimentation. As the organization freezes, employee engagement and morale often plummet. The rigid environment and lack of opportunities for growth or creative input make employees feel undervalued and disconnected from the company's mission. The work becomes monotonous and employees may become disengaged, leading to lower productivity, increased absenteeism, and higher turnover rates. The company's culture suffers as a result, further entrenching the frozen state.

A frozen system is often characterized by an inflexible organizational structure that doesn't align to the needs of the business for quick adaptation to new challenges. Hierarchical and siloed departments make cross-functional collaboration difficult, if not impossible. The rigid structure prevents the company from reallocating resources or reconfiguring teams in response to changing business needs. This inflexibility can lead to missed opportunities and an inability to compete effectively in a dynamic market.

In a frozen system, decision-making is often concentrated at the top levels of the organization, with little empowerment of lower-level managers or employees. This centralization leads to bottlenecks, as every significant decision must pass through a few key individuals who may be disconnected from the day-to-day operations. The result is slower response times, frustration among employees who feel their input is ignored, and a general lack of agility in the company's operations. The frozen state isn't exclusive to large companies; even small startups can experience it. Imagine a team of just two to five people led by an overzealous CEO who has strong opinions and rigid judgments about everyone's work. While the simplicity of a small team might initially mask the impact of this behavior, it's a ticking time bomb.

As the business grows and complexity naturally increases, the CEO's need for control will stifle progress, which locks the team into a frozen state where decisions slow, autonomy disappears, and the organization becomes unable to adapt. Even at this scale, the frozen state can take hold, halting momentum before it truly begins.

To prevent a team or a company from entering a frozen state, it's essential for leaders to cultivate a culture of adaptability, where change is embraced and encouraged. This can be achieved by streamlining processes, reducing unnecessary bureaucracy, and promoting an environment that values innovation and employee input.

Leaders should also work to decentralize decision-making, empowering employees at all levels to contribute to the company's direction and success. By maintaining flexibility and openness to new ideas, the organization can avoid the pitfalls of becoming rigid and unresponsive, ensuring continued growth and relevance in a competitive market. When a company strikes this balance, it reaches an optimal state: flowing.

Flowing: Stability and Flexibility

The ideal state of "flowing" resembles the characteristics of a liquid state. Just as water flows smoothly and adapts to its environment, a company in the flowing state operates with agility, adaptability, and efficiency. This is the sweet spot where the organization is neither chaotic (boiling) nor rigid (freezing) but instead maintains a dynamic equilibrium that allows for continuous growth and innovation.

In a flowing state, the company is able to respond quickly to changes in the market, customer needs, and internal challenges. This agility comes from a well-structured yet flexible system where information flows freely between departments and decision-making is both rapid and informed. Teams are empowered to make decisions at the appropriate levels, reducing bottlenecks and ensuring that the organization can pivot or scale as needed. The ability to adjust strategies and operations on the fly can be a key competitive advantage.

Flowing organizations have mastered the art of efficient collaboration and communication. Just as liquid molecules move smoothly past one another, departments and teams within the company interact seamlessly. Clear channels of communication and well-defined processes prevent misunderstandings and delays.

Regular feedback loops ensure that all parts of the organization are aligned with the company's goals. Any issues are quickly identified and addressed. This state of flow is characterized by a high level of trust and coordination among employees, which encourages a strong sense of shared purpose.

A business in a flowing state is constantly evolving. Continuous improvement is embedded naturally in the culture, with regular retrospectives and process reviews leading to incremental innovations and optimizations. The company's systems and processes are designed to be flexible and scalable.

Accountabilities and responsibilities are clear for everyone. The focus on continuous growth ensures that the organization remains competitive and capable of handling increasing complexity without slipping into chaos or rigidity.

In a flowing state, there's a delicate balance between stability and flexibility. The company has established enough structure to provide direction and stability, but not so much that it stifles creativity or responsiveness.

This balance allows the organization to maintain order while still being able to innovate and adapt. Just as liquid water is stable yet capable of moving and changing shape, a flowing business is resilient, able to absorb shocks and adjust course without losing its momentum.

Let's take a look at how these new concepts affect Alice and Theo's experience.

Alice and Theo: The Path Forward

Theo's hypothesis that he's just doing more work is partially correct, but he doesn't realize that the ad hoc nature of their current processes is becoming unsustainable. Without proper structure, they're approaching a technical boiling point where the system can no longer be efficiently managed.

This chaotic state can lead to bugs slipping through, decreased productivity, and, ultimately, a loss of control over the system. Alice and Theo need to determine the best time to strengthen their process before the boiling point is reached.

If they wait too long and the system boils over, more radical intervention will be needed to pull them back into a flowing state.

Alice and Theo's software has gone through an invisible phase change. Initially, Theo spent most of his time writing new code, but now, he finds himself reading more code than he writes. This shift is a common sign of growing technical complexity, but because they have no clear way to measure it, they aren't fully aware of what's happening.

Theo feels the strain, but he struggles to articulate exactly why things are slowing down. He attributes it to an increased workload, but the real problem is deeper—technical complexity is accumulating in a way that makes each new change harder to implement. Alice, looking for a way to ease Theo's burden, might consider hiring more developers.

However, this approach won't solve the underlying issue. Adding more people to the team without addressing the root cause will likely worsen the problem, as additional developers will also spend more time navigating the growing complexity before they can contribute effectively. The slowdown will compound as more code requires more manual testing, making it harder for the team to move forward at the same pace.

Instead of charging ahead with a quick fix, Alice and Theo need to pause and reassess their approach. They don't yet have a clear plan, but they recognize that continuing as they are will only make things worse. Their challenge is no longer just delivering features—it's ensuring that the foundation of their system can support future growth.

Alice Understands Theo's Situation

After a break from their initial conversation, Alice and Theo sit back down to discuss the issue further.

Alice begins, "Theo, I want to make sure we're both on the same page. It's really important to me to understand what's going on for you so we can address the situation properly. I need to know how we can support you and improve our processes."

Theo appreciates Alice's willingness to listen. "I know it might seem like things are just getting too complicated, or that I might be losing interest, but that's not the case. I'm still fully committed to this project. The growing complexity of the codebase is just making everything take longer. Since there's more code now, each feature takes longer to test and more effort to release. I can't keep doing these tasks manually if we want to move faster."

Alice nods, her expression earnest. "I realize that now. It's not about your skills or dedication. It's about the systems we have in place. The more we build, the harder it is to manage everything manually. I want to find a way to help you manage this workload better because I can see how much effort you're putting in."

"That's a big part of it, yes," Theo says. "Automated tests could catch issues faster and more reliably than manual tests. And if we automate the deployment process, we can save a lot of time and reduce human errors. I can work on all of these things to speed my work back up, but if I do, I'm not working on the features our customers want now."

Alice begins to see the bigger picture. "Alright, I think I understand now. We've been so focused on shipping features quickly that we haven't built the support systems needed to keep that pace sustainable. What would we need to do to implement these changes?"

Theo explains, "First, we should add more automated tests. Automated tests run with every change to the codebase, ensuring new features don't break existing functionality. They're faster and more reliable than manual tests. If I spent a little more time with each feature I build to add more automated tests, we would get much further ahead."

"That sounds like a good start," Alice says. "What else?"

"Second, we should automate the deployment process using continuous integration and continuous deployment (CI/CD) tools. CI/CD would automatically build, test, and deploy our code whenever we make changes, which reduces the time spent on these tasks and minimizes the risk of human error. Instead of me having to manually run every step of the deployment process, I could just push a button and the tool would do it automatically."

Alice is beginning to feel hopeful but also remains deeply concerned. "Wow! I would love for you to just push a button. But that seems like it will take a long time to build, and we need to keep getting features out the door."

"Do we have any budget to bring in a DevOps engineer? Even if they're part time?" Theo asks. "A DevOps engineer could set up and maintain the CI/CD pipelines, help me integrate the automated tests into the pipeline, and ensure that everything runs smoothly."

Alice nods. "Alright, Theo, let's move forward with this."

Liquid

Chapter 12:

Reducing Complexity

Alice and Theo's challenges mirror a larger trend in the industry. As our software products continue to grow, we face a critical challenge: How do we continue delivering value at a consistent rate as the surface area of the project gets larger?

One common misconception is that simply adding more developers to the team will solve the problem. Yet this simplistic attempt to solve a complex situation is the path that most companies choose to take.

Unfortunately, blindly hiring more people can exacerbate the issue rather than alleviate it.

The Pitfalls of Adding More People

Adding more people to a project might seem like an intuitive solution to speed up development and manage increased workloads. However, in a complex system, this approach often leads to unintended consequences.

Difficulty Explaining the Codebase

Each new developer needs to understand the existing codebase before they can contribute effectively. This involves familiarizing themselves with the overall architecture of the system, the design patterns used, and the specific coding conventions followed.

They need to grasp the logic behind existing functionalities, understand how different modules interact, and become aware of any known issues or technical debt within the system. This onboarding process requires significant time and effort from current team members, diverting them from their core tasks. The more complex the system, the longer it takes for new hires to become productive.

No Reduction in Complexity

Hiring more developers doesn't address the root cause of the slowdown, which is the increasing complexity of the codebase. Without efforts to refactor and simplify the code, new developers will spend much of their time navigating convoluted structures, further slowing down progress. Each piece of new code can introduce additional dependencies and interactions that further complicate the system.

Over time, the codebase can become a tangled web of interdependent modules, making it difficult to implement changes or add new features without introducing bugs. This added complexity can result in more time spent debugging and less time developing new features, reducing overall productivity.

The cognitive load on each developer increases as the codebase grows more complex. Developers must keep track of numerous variables, functions, and interactions, which can lead to mental fatigue and errors. Complex codebases also make it harder to ensure code quality, as thorough reviews become more challenging and time-consuming.

Increased Burn Rate Without Corresponding Value

Hiring more employees leads to higher operational costs, including salaries, benefits, and additional overhead expenses. This accelerates the company's burn rate—the rate at which it spends capital to sustain operations—without necessarily delivering proportional value, increasing financial pressure and reducing runway.

If the added headcount doesn't result in proportional productivity gains, the company may face financial strain without realizing significant improvements.

Heightened Stress

As complexity grows and deadlines loom, the stress levels of the entire team can rise. Current developers may feel overwhelmed by the need to support and train new hires while also managing their own tasks and keeping up the pace of development. This can lead to burnout, reduced morale, and decreased overall productivity.

Strategic Approaches to Reducing Technical Complexity

To encourage sustainable growth, we must focus on reducing the system's complexity. Here are some strategies to achieve this.

Refactoring

When we recognize that technical complexity is growing, our first step should be to prioritize refactoring the codebase to streamline and simplify it.

Refactoring is a critical process in managing and reducing technical complexity. It involves restructuring existing code without changing its external behavior. This practice addresses the technical debt accumulated over time and enhances the system's maintainability and scalability.

Paying off technical debt through refactoring creates new boundaries in the software. These boundaries help compartmentalize the system, ensuring that there's less information inside each boundary than outside. By breaking existing entities into new sub-entities, developers can isolate different parts of the code, making it easier to understand and manage. This modular approach allows for more straightforward maintenance and modification, as changes in one part of the code are less likely to affect other parts.

Refactoring reduces complexity by simplifying the code structure. As developers streamline the code, they eliminate redundancies and improve readability. This often results in shorter programs that are easier to navigate and understand. By reducing the cognitive load on developers, refactoring enables them to focus more on developing new features rather than getting bogged down by the intricacies of a convoluted codebase.

When we refactor our codebases, we enhance the system's robustness and efficiency. A cleaner, more organized codebase will facilitate quicker and more reliable feature development, ensuring that the team can continue to innovate and meet user demands without being hindered by technical debt.

Automating Processes

Implementing automated testing and CI/CD pipelines can significantly reduce the manual workload for development teams. Automated tests catch issues early, ensuring new features don't break existing functionality. By running these tests continuously, developers receive immediate feedback, allowing them to address problems quickly.

Automated testing encompasses unit tests, integration tests, and end-to-end tests, which collectively enhance software reliability and free up developers from manual testing tasks.

- **Continuous integration (CI)** involves frequently merging code changes into a shared repository, automatically building and testing the codebase with every change. This practice helps detect integration issues early, reducing the time needed to resolve conflicts and bugs.

- **Continuous deployment (CD)** automates the deployment of code to production, minimizing human error and ensuring smooth and consistent releases. This process enables faster release cycles, allowing the team to deliver updates and improvements to users more frequently.

Together, CI/CD pipelines streamline the workflow, enhancing the speed and quality of software releases while reducing the manual workload. For Alice and Theo, adopting CI/CD will be crucial in managing their growing complexity and maintaining a competitive edge in the market.

Let's do some math on the real costs to an organization from not automating:

- H = hours per week a developer spends on manual testing or manual deployment
- C = average cost per hour for a developer
- N = number of developers in the organization
- W = working weeks per year per developer (e.g., 50 weeks)

Annual lost productivity due to inadequate automation = H * C * N * W

For example:

- H = 5 hours per week spent on manual testing
- C = $100 average cost of each developer per hour
- N = 10 engineers on the team
- W = 50 working weeks per year

Annual lost productivity due to inadequate automation in this example is projected at 5 * $100 * 10 * 50 = $250,000

For the same cost as the $250k lost annually to productivity inefficiencies, this company could hire one or two developers dedicated entirely to building and maintaining automation systems.

These automation-focused roles would drastically reduce manual tasks like repetitive testing and deployments, allowing feature developers to concentrate on delivering value to customers. By streamlining processes through automated testing and CI/CD pipelines, the overall efficiency of the development team would improve, reducing bottlenecks and accelerating feature delivery.

Investing in automation isn't just about recovering lost productivity. It's about creating a scalable foundation for growth. Automation reduces the likelihood of errors, ensures faster feedback cycles, and minimizes the cognitive load on developers who would otherwise be tied up with manual processes.

This means the entire team can spend more time on innovative features, strategic problem-solving, and activities that directly impact the company's bottom line, effectively turning a sunk cost into a strategic investment.

Focused Hiring

Instead of hiring a large number of generalized developers, it would be more effective to prioritize specialized roles that currently lack dedicated support.

Alice and Theo have decided to bring in a specialized DevOps engineer to address a critical gap in their company's infrastructure. This role will be responsible for setting up and maintaining automation tools, optimizing infrastructure, and streamlining development processes, allowing the team to manage complexity more effectively.

Since there's no existing infrastructure in place for this role, the new hire can hit the ground running with standard implementation practices, quickly establishing foundational systems that improve efficiency.

Beyond the immediate improvements, hiring a specialist also creates lasting benefits for the company. A well-structured DevOps function reduces deployment friction, minimizes downtime, and enhances system reliability, allowing developers to focus more on feature development rather than firefighting operational issues. Over time, this investment leads to a more scalable, resilient engineering organization that can sustain growth without becoming overwhelmed by technical bottlenecks.

By addressing the underlying complexity issues and implementing strategic improvements, a company can enhance their team's productivity and create a solid foundation for sustainable growth. Reducing complexity not only makes the current system more manageable but also prepares the company to scale effectively as it continues to expand. Ultimately, when determining how to resolve technical complexity within a system, technology executives should focus on resolving this situation in the following order:

1. Automating areas where recurring manual work is common
2. Refactoring code to create better boundaries
3. Hiring into specialized roles where they have gaps and/or little infrastructure
4. When the above areas have been addressed, scaling in more feature developers

Chapter 13:

Self-Organization— Empowering Teams and Processes to Adapt

By resolving technical complexity, teams are able to focus on higher-level problem-solving and organizational alignment. One of the most remarkable traits of complex adaptive systems (CAS) is their capacity for self-organization. If the conditions are right, systems lean toward arranging their components in purposeful and efficient ways without external control.

In the context of a business, this means that teams, processes, and even entire departments can align themselves to meet goals, respond to challenges, and adapt to changing circumstances when the right environment is cultivated.

The Mechanics of Self-Organization

Self-organization emerges when individuals within a system are empowered to make decisions within clear boundaries and with access to shared information. It's not about chaos or complete autonomy; rather, it's about setting the boundaries that create conditions where people can independently align their actions within the context of the larger goals of the organization.

Imagine a product team tasked with delivering a new feature. In a rigid, overly structured system, every decision might need approval from a manager, causing delays and stifling creativity. You might recall that we refer to this system as frozen. In a self-organizing system, however, the team understands the company's strategic objectives, has access to shared resources, and can collaborate to decide the best course of action without constant oversight. They consult their manager when they need to and keep them informed. The result is faster decision-making, more innovation, and greater adaptability.

The Role of Leadership in Self-Organization

Self-organization doesn't happen in a vacuum; it requires intentional effort from leadership to create the conditions that allow it to thrive, focusing on clear boundaries, transparency, and trust.

Clear Boundaries

Boundaries are essential for self-organization, providing the framework within which teams can operate effectively. Leaders must define the scope of autonomy for teams, ensuring they understand where they have the freedom to make decisions and where alignment with broader organizational goals is required.

For instance, an engineering team might be empowered to choose their own tools and frameworks for development but must ensure that their choices align with the company's overarching technology stack and meet established budgetary, security, and compliance standards.

These boundaries prevent teams from veering off course while still allowing them to innovate and act decisively within their domain. Boundaries, however, must also flow upward, guiding how teams interact with leadership. Teams need clarity on when to escalate decisions, seek approval, or provide updates to ensure alignment at higher levels. For example, a product development team might be autonomous in sprint planning but required to present monthly updates on their roadmap to leadership. This upward flow of boundaries ensures that leaders remain informed and can intervene when necessary without micromanaging daily activities.

Balancing these upward and downward boundaries is critical to maintaining a healthy, adaptable system. Too much restriction stifles creativity and creates bottlenecks, leading to a frozen state, while too little structure results in chaos and inefficiency, akin to a boiling system. Clear, reciprocal boundaries allow teams to operate with confidence and autonomy while ensuring the overall system remains aligned and cohesive, enabling sustainable growth and adaptability.

Transparency

Transparency is essential for self-organization, but it must flow both downward and upward within the organization. Leaders need to ensure teams have access to critical information, such as company goals, performance metrics, and customer feedback, so they can make informed decisions that align with the broader strategy. Without this clarity, teams are left to operate in silos, disconnected from the company's larger objectives.

Equally important is promoting a culture where teams are transparent with leadership. This involves encouraging teams to regularly share updates, provide metrics on their progress, and document their work through tools like product briefs or project reports. When teams are open about their challenges, achievements, and decisions, it enables leaders to offer meaningful guidance, address systemic issues, and ensure the organization remains aligned.

Transparency in both directions not only improves decision-making but also builds mutual trust. Leaders who provide visibility into the "why" behind strategic goals empower teams to act autonomously. Teams that communicate openly about the "how" of their execution ensure leadership has a clear understanding of what's happening across the organization. This reciprocal flow of information strengthens the system as a whole, enabling greater cohesion, adaptability, and long-term success.

Trust

Self-organization thrives in a culture where trust freely flows both downward and upward within the organization. Employees must feel trusted to make decisions without being micromanaged, knowing that their leaders believe in their capability to act in the company's best interest. This empowers teams to take ownership of their responsibilities, promoting innovation and agility. At the same time, leaders must act in a manner that earns trust from their teams and peers, demonstrating transparency, reliability, and alignment with the organization's goals.

When employees see their leaders modeling accountability, openness, and consistency, it reinforces their willingness to embrace autonomy while staying aligned with the broader mission. Trust is a two-way street, creating a feedback loop where both leaders and teams support one another in navigating the complexities of the system.

Self-Organization in Action

Self-organization is particularly evident in teams that embrace **lean and iterative workflows**. These teams operate with a clear understanding of their objectives and are empowered to make decisions that align with the company's broader goals. Rather than waiting for top-down directives, they adapt and respond dynamically to challenges, ensuring continuous progress.

Consider our earlier example of a software development team tasked with delivering a new feature. Instead of rigidly following a predefined plan, the team organizes itself to work in smaller iterations, delivering incremental improvements while continuously testing and refining their work.

Each team member contributes their expertise, and together, they decide how best to approach the problem. This iterative process enables them to deliver value to the customer faster while maintaining the flexibility to pivot when necessary.

Or let's consider a platform team focused on optimizing infrastructure. With the autonomy to prioritize their work, they identify bottlenecks and proactively implement solutions that improve system performance. Their ability to self-organize means they can address technical debt or scalability issues without waiting for explicit instructions, ensuring the system remains resilient and adaptable.

Self-organization isn't about operating without oversight but rather about teams having the freedom to act within clear, well-defined boundaries. As you can see from these examples, by fostering a culture of trust, providing transparency, and setting the right boundaries, leaders can create the conditions for teams to self-organize effectively. This will not only drive innovation but also maintain liquid flow throughout the organization.

The Benefits of Self-Organization

Self-organization helps businesses maintain a liquid state, where the system flows smoothly and remains adaptable. Instead of relying on top-down directives for every decision, self-organizing teams take ownership of their work, aligning with company goals while dynamically adjusting to challenges.

A self-organizing team is capable of responding quickly to market shifts. If a customer needs a change, the team can adapt without waiting for leadership approval at every step. For example, a product team might recognize that users are struggling with a particular feature and proactively experiment with improvements, deploying a fix in days rather than weeks. Similarly, an engineering team can detect inefficiencies in their development pipeline and introduce automation or workflow adjustments without needing executive intervention.

This level of adaptability not only improves responsiveness but also encourages innovation. When teams feel empowered to make decisions, they're more likely to experiment with new ideas, iterate processes, and find creative solutions to problems. Companies like Spotify and Netflix have famously embraced self-organization, using small, autonomous teams that take ownership of features, experiment rapidly, and drive continuous innovation—all without excessive bureaucracy slowing them down.

Beyond improving agility and innovation, self-organization also reduces the load on leadership. When teams can handle decision-making at the appropriate level, executives and managers are free to focus on high-level strategy and systemic improvements rather than getting caught up in day-to-day execution.

Instead of constantly firefighting, leaders can work on influencing the system as a whole, setting the right guardrails and direction while trusting teams to operate within them. However, self-organization doesn't mean a lack of structure, direction, or oversight. Teams still need clear goals and boundaries from leadership to avoid drift, ensure alignment with broader company goals, and prevent fragmentation. Without this guidance, teams may optimize for local efficiencies that unintentionally create silos or misalignment across the organization.

Leaders play a crucial role in balancing autonomy with accountability, ensuring self-organization contributes to the overall health of the system. This entails setting shared priorities, defining key outcomes, and establishing principles that guide decision-making while still allowing teams the flexibility to determine how best to achieve those outcomes.

For example, a company might establish engineering standards, security requirements, or customer experience principles that all teams must follow, ensuring consistency and cohesion while still enabling autonomy. When implemented effectively, this framework creates a balance between freedom and alignment, allowing teams to self-organize while ensuring the business moves forward as a unified system.

Avoiding Pitfalls

For self-organization to thrive, organizations must provide clear goals, aligned incentives, and the right level of autonomy. Without these, teams may drift in different directions or struggle with misalignment. However, when structured properly, self-organization creates an environment where businesses can scale efficiently, remain resilient in the face of change, and sustain long-term growth.

Remember that one of the key emergent properties of a liquid state encourages Speed. For speed to exist, self-organization is a critical tool for reducing drag and maintaining momentum. By empowering teams to adapt and align themselves, businesses can accelerate their growth while ensuring the system remains stable and efficient.

Chapter 14:
The Path Forward—Aligning Vision and Execution

For Alice and Theo to make the strategic decision to focus on reducing complexity, their conversation needed to first be grounded in mutual understanding and clear communication.

Theo, being a savvy software engineer, understood that as their software adapted to the product space, it was crucial to pay down technical debt. He knew this intellectually, but he was also under pressure to push out new features to satisfy the growing user base.

Theo didn't recognize that *technical complexity was growing within the overall system*, but he did recognize a clear action plan. Theo realized he needed to add more automation to reduce his manual tasks. The outcome of this would be gaining that time back to focus on features.

Theo is beginning to develop a key skill that's present in systems thinkers—he's beginning to see the downstream impacts to the overall system two, three, or even four levels of abstraction away from the action being taken. While Theo doesn't yet consciously "see the system," he's starting to learn how to productively influence it.

There is always a complex adaptive system (CAS) at work within your organization. You don't have to see it or create it. It's simply always there.

However, Theo now needs to convey this to Alice, who's been focused on the immediate needs of growing revenue and user acquisition through continuous feature development. Alice is concerned about being able to keep up with customer demand for the key necessary features so they can continue to grow their customer base and their revenue.

In their conversation, Theo needs to be able to explain that while pushing out new features is important, doing so without addressing the growing complexity will lead to slower development, more bugs, and higher costs in the long run. He could explain that paying off technical debt involves creating boundaries within the software, where there's less information inside each boundary than outside.

He needs to be able to explain that by breaking entities into sub-entities, they would reduce the complexity, making the codebase more manageable and efficient—which, in turn, would increase the speed of core feature development.

Despite Theo's technical insights, he needs to communicate these concepts in a way Alice will understand. Alice might initially respond with concern, focusing on the immediate need to grow their revenue and user base.

Theo could frame the conversation around the long-term benefits: "Alice, if we continue to build on a complex system, we'll face increasing development times, more bugs, and higher costs. By taking time now to reduce complexity, we're investing in faster feature development, better product stability, and lower costs in the future."

Theo could also use analogies to make his point clearer. For instance, he might say, "Imagine our software is like a city. Right now, it's growing without a plan, and the roads are getting tangled. We need to build proper roads and infrastructure before adding more buildings.

It might slow us down a bit now, but it will save us time and money in the long run." While Theo should speak in less technical terms and use relatable analogies, it's equally important for Alice to actively listen and see to understand the root cause of the slowdown. Together, these strategies will ensure Alice gains understanding of how the investment in reducing complexity is crucial for sustainable growth.

By aligning their visions and focusing on both immediate needs and long-term stability, Alice and Theo can now move forward with a balanced approach, ensuring their company's continued success.

The Real Cost of not Seeing Complexity

Many people operating within a CAS are unaware of its existence. They can ignore it or pretend it doesn't exist, but the system will continue to operate beneath the surface. The system is always there, even if you aren't aware of it.

When a CAS isn't functioning properly, everyone within it experiences stress. Theo, for example, isn't a bad person, and he's not poor at his job either. He's fully committed and doing his best to ensure the business grows. However, he's getting frustrated, and so is Alice.

This confusion is scary, and when people are stressed, confused, and fear that their company will fail, they'll try anything to get it back on track. *People will begin to devolve to their worst selves when experiencing this level of stress.*

When leaders believe their untrue hypotheses without examining the underlying system, they can cause significant, lasting damage to their company. When a company exists in a continually frozen or boiling state for a long time:

- People become unhappy, which leads to the formation of negative cultural norms.

- Trust erodes between individual contributors, teams, and leadership.

- A lot of action takes place to cover up symptoms, but no actual progress is being made.

Most leaders who don't understand the underlying system at work will focus purely on trying to fix symptoms. For example, requiring all engineers to return to the office to collaborate more in hopes of increasing productivity addresses the symptom only. Why do you need them to collaborate more? Are they truly not collaborating, or is there underlying complexity at play? What's the root cause behind the lack of collaboration? What's percolating consistently within your system that you really need to address?

Assigning a cross-functional team to tackle this complexity would be a more effective solution. By creating a cross-functional team, you're creating a new boundary within your system and a new directed set of interactions, which will help you simplify and focus on an effective solution. Bring only this team into the office so that they can fast-track their communications, for a specific purpose, within the boundaries of this team.

By recognizing and understanding the complex system at work, leaders can move beyond addressing superficial symptoms and start tackling the true core problems. This shift from spontaneous reactions to a deeper curiosity about the secret world of CAS can lead to more sustainable and effective solutions, reducing stress and confusion within the organization.

Managing Boundaries Effectively

Throughout this book, we've explored the concept of creating new boundaries within a company's system to manage complexity and prevent chaos.

In a complex system, a boundary defines the scope of interaction and influence. It helps shape the system's behavior by constraining interactions and focusing adaptation within specific limits.

Guiding a Boiling System Toward Flow

In a scaling company, clear boundaries are essential because they help to organize and contain the interdependencies between various elements of the system. When our business begins to boil, becoming chaotic and overwhelmed by increasing complexity, we can help settle it back into a manageable, flowing state by introducing new boundaries.

In a boiling business, the interactions between teams, processes, or projects become too entangled, leading to inefficiencies, confusion, and often erratic performance. When the system reaches this point, it can feel as though every decision, action, or change triggers a cascade of unintended consequences.

Clearer boundaries compartmentalize the complexity and make it easier to manage. These boundaries can take the form of well-defined roles, more structured processes, or even a physical or digital separation between different teams or systems within the company.

For example, a growing organization that struggles with overlapping responsibilities between teams lacks clear boundaries, which can cause duplication of work and misaligned priorities. By establishing defined areas of ownership and responsibility for each team, essentially creating boundaries between them, leaders can reduce friction and confusion, allowing teams to focus more effectively on their specific goals. These boundaries help limit unnecessary cross-team dependencies, creating a sense of clarity and order within the company.

Once these boundaries are in place, the system stabilizes, transitioning from a chaotic, boiling state to a more liquid, flowing state where the company operates with greater agility and coordination. These boundaries don't restrict movement; rather, they create a structured environment where teams can collaborate efficiently without stepping on one another's toes.

Thoughtfully implemented boundaries help define the scope of interactions, ensuring each team or subsystem operates with clarity and focus.

This reduces unnecessary overlap, minimizes miscommunications, and provides a framework that supports decision-making. As the system moves into flow, decision-making becomes clearer, feedback loops become more manageable, and overall productivity increases. With well-designed boundaries, teams can operate autonomously while staying aligned with broader organizational goals, promoting both efficiency and innovation.

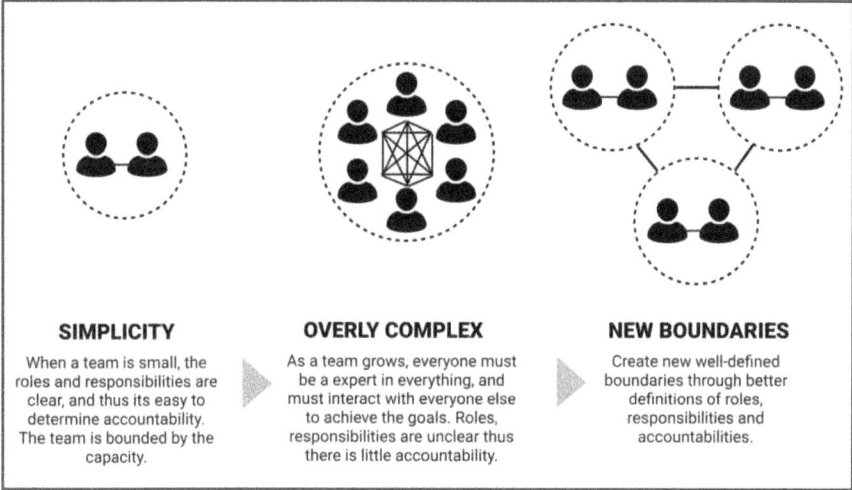

SIMPLICITY	OVERLY COMPLEX	NEW BOUNDARIES
When a team is small, the roles and responsibilities are clear, and thus its easy to determine accountability. The team is bounded by the capacity.	As a team grows, everyone must be a expert in everything, and must interact with everyone else to achieve the goals. Roles, responsibilities are unclear thus there is little accountability.	Create new well-defined boundaries through better definitions of roles, responsibilities and accountabilities.

This structured approach prevents the system from devolving into chaos by managing complexity before it spirals out of control. Instead of reacting to crises or being overwhelmed by boiling chaos, leaders proactively shape the system to maintain liquid equilibrium.

Boundaries create predictable patterns of interaction, which serve as a foundation for sustained performance and allow the organization to respond effectively to new challenges and opportunities.

Guiding a Frozen System Toward Flow

While a boiling system struggles with chaos and overstimulation, a frozen system is burdened by rigidity and resistance to change. In this state, teams often find themselves mired in excessive bureaucracy, reluctant to innovate, and constrained by outdated processes. The result is an organization that struggles to adapt, with progress stifled and opportunities missed.

Guiding a frozen system toward flow requires the deliberate introduction of flexibility, trust, and collaboration, which restores the organization's capacity for movement and adaptability.

Frozen systems reveal themselves through bottlenecks, where teams are paralyzed by waiting for approvals or overwhelmed by the fear of making mistakes. In these environments, "the way it's always been done" becomes a mantra that inhibits creative problem-solving. Departments operate in isolation, often duplicating efforts or failing to share critical knowledge. Processes grow heavier with time, creating a culture where even minor adjustments require navigating a labyrinth of permissions and protocols.

To help a frozen system thaw, leaders must simplify and energize their organizations—not by removing boundaries but rather by redefining and recalibrating them. One of the first steps is to assess the processes that govern daily operations. Many of these processes, while originally designed to create order, may now hinder progress. By eliminating unnecessary steps or consolidating workflows, teams can begin to move more freely.

Empowering team members with decision-making authority within well-defined boundaries fosters a sense of autonomy and confidence, enabling them to act quickly and decisively without seeking constant validation from higher-ups.

Collaboration is another essential ingredient in melting a frozen system. Silos within an organization contribute to stagnation, with teams operating in isolation and failing to leverage the strengths and insights of their peers. Creating opportunities for cross-functional collaboration, such as joint planning sessions or regular retrospectives, allows teams to align on goals, share ideas, and spark innovation. As trust and communication improve, the system begins to flow more naturally.

Equally important is cultivating a culture that challenges the status quo. Leadership plays a pivotal role in this transformation by modeling curiosity and openness. When leaders actively question outdated practices and highlight successful examples of change, they inspire their teams to embrace experimentation and growth. Storytelling can be a powerful tool in this context, showcasing how adaptive thinking has led to meaningful results in other parts of the organization.

The transition from a frozen state to one of flow requires careful, sustained effort. As rigidity gives way to flexibility and collaboration, the organization regains its ability to respond dynamically to challenges and opportunities. Leaders must continue to nurture this adaptability, balancing structure with freedom, so the system remains fluid and aligned with its goals. In the end, a once-frozen organization becomes one that's alive with possibility, moving steadily and purposefully toward success. When leaders succeed in establishing these boundaries, they guide the organization into a sustained flow state. The business is no longer negatively impacted by inefficiencies, as teams can operate with clarity, minimizing friction and redundant efforts. Decision-making becomes more streamlined, reducing bottlenecks and allowing work to progress smoothly. Collaboration improves as teams understand their roles and responsibilities within the system, encouraging innovation and accountability.

Chapter 15:

Moving From Reactive to Proactive

A common misconception about complex adaptive systems (CAS) is that they're a method for building an organization from scratch. Leaders don't create the CAS; they can only influence it. The system is always there, operating beneath the surface, and our role is to observe and understand its dynamics so we can influence it to change in the direction that we desire.

To navigate the challenges within a CAS, leaders must adopt an ontological pattern—a guiding framework that illuminates how to manage complexity as the system evolves and gains new capabilities. By cultivating the ability to deeply observe the system's dynamics, leaders can identify the areas that are locked in rigidity or spiraling into chaos. With this understanding, they can apply deliberate interventions to restore balance, guiding the system toward a state of fluid and sustainable flow.

When your CAS starts to wobble, it's crucial to identify the areas that are boiling (chaotic) or frozen (overregulated), then influence the system to move toward correcting this. The longer the CEO, CTO, and all other leaders within the company wait to correct this wobble, the more difficult coming back to liquid will be. If you're an experienced systems thinker, you would proactively work to identify potential boiling or freezing points within the system and build a list of potential phase change strategies. By identifying these phase changes as they begin to happen, the team can build a cadence to take down the complexity before the phase change negatively affects the organization.

For instance, Alice and Theo could have realized that they would be at risk of freezing while rushing to their MVP if they had recognized certain warning signs. Delays in decision-making, where progress relied too heavily on Theo's approval, could have indicated a bottleneck forming. Similarly, if key individuals, such as Theo, were juggling too many responsibilities, from coding to strategic planning, it would have pointed to an overreliance on specific team members—a common precursor to freezing. They might also have observed a reluctance to iterate or launch imperfectly, reflecting an overly rigid mindset focused on perfection rather than adaptability.

Inflexible processes, such as requiring multiple layers of approval for every change, and silos between product, development, and feedback loops could have been additional red flags. These conditions often lead to inefficiency and a lack of responsiveness, hallmarks of a frozen system. By building slack into their schedules, decentralizing decision-making, and encouraging collaboration between teams, Alice and Theo could have mitigated the risk, maintaining the flexibility needed to approach their MVP with adaptability and flow. Recognizing and addressing these signals early would have helped them avoid unnecessary stagnation during a critical phase of their business.

With this foresight, they would have understood that there would be a period shortly after producing the MVP where they would need to reduce complexity before moving on to new feature development.

However, because they were blind to the underlying system at work, they had to adapt reactively after the consequences were already being felt.

If we want to be effective leaders, it's our responsibility to recognize the conditions under which these phase changes will start to happen, then proactively put the processes in place to encourage the system back toward simplicity before it topples over into the frozen or boiling state.

Planning for this pause to reduce complexity is essential because it allows the system to remain in a flow state. By anticipating and planning for the system's natural tendencies to become either chaotic or overregulated, leaders can maintain a balance and ensure sustainable growth.

Recognizing that we're part of a complex system, and that our role is to influence it rather than control it, is a cornerstone of effective leadership and management. This awareness shifts our approach from attempting to dictate outcomes to understanding how to guide the system toward desired behaviors and goals.

When You're Boiling and Frozen

It's quite common for parts of your company to be frozen while others are boiling. This duality occurs because different teams, departments, or processes within the organization experience varying types of complexity at different times. For example, your sales team might be operating in a boiling state, overwhelmed by too many conflicting priorities and customer demands, leading to chaos and inefficiency. At the same time, your product development team could be frozen, bogged down by overly rigid processes or outdated systems that stifle creativity and adaptability. These contrasting states can create significant friction across the organization, making it difficult to align on goals or move forward cohesively.

This mismatch often exacerbates systemic issues. Boiling teams may overwhelm frozen teams with demands they're not equipped to handle, leading to frustration and strained relationships. Conversely, frozen teams may slow down boiling teams by introducing bottlenecks or requiring excessive approvals, further amplifying the chaos.

Recognizing and addressing these imbalances is critical to maintaining organizational flow. Leaders must identify which parts of the company are boiling and which are frozen, then tailor their interventions to bring the entire system back to a liquid state. By doing so, they create an environment where all parts of the organization can work together harmoniously, supporting each other's goals and driving the company forward as a unified flow.

Chapter 16:
The Role of Empathy and Active Listening

Empathy and active listening are critical skills in navigating and managing the complex adaptive systems (CAS) within our organizations. By truly listening and empathizing, we gain access to deeper perspectives, uncovering the challenges and subtle dynamics that shape individual roles within the system. Instead of only seeing symptoms, we begin to recognize the underlying blockages preventing the system from flowing productively. This shift is essential for comprehending the full scope of what's happening within our business beyond just the visible problems.

Leadership isn't about issuing mandates to address surface-level issues. It requires digging deeper, understanding root causes, and taking thoughtful, deliberate action to resolve them. Without empathy and active listening, leaders risk becoming part of the very problem they seek to fix, trapped in a cycle of reacting to symptoms instead of addressing the forces driving systemic dysfunction.

True leadership begins with the courage to ask hard questions: What's really causing these challenges? Where are the systemic failures? This mindset allows us to see the system for what it truly is—a complex network of interactions that, when understood, holds the key to long-term success. Once you can see the system clearly, you gain the ability to influence it intentionally.

This *shift from reactive problem-solving to proactive system management is transformative.* By aligning actions with the deeper dynamics at play, leaders create the conditions for sustainable success, ensuring the organization operates with resilience and purpose rather than just reacting to crises.

Who and Why

In a CAS, leaders must listen to and empathize with everyone involved—developers, managers, stakeholders, and customers. Each group experiences the system differently, and their insights are invaluable for understanding its intricacies. Recognizing that many challenges stem from the nature of a complex system rather than individual shortcomings shifts our perspective, allowing us to focus on systemic barriers instead of assigning blame.

Compassion and curiosity go hand in hand. By actively listening and seeking to understand what's happening beneath the surface, we uncover root causes rather than react to symptoms. This deeper insight ensures we solve the right problems—ones that lead to meaningful, long-term change rather than quick fixes that fail to address the core issues. Empathy, in this way, becomes a strategic tool for building a more resilient and effective organization.

For example, instead of viewing Theo as disengaged or unmotivated, Alice now sees his frustrations as a reflection of increasing technical complexity.

This realization helps her recognize her own role in the situation. By shifting her perspective from blame to understanding, Alice creates an opportunity for collaboration. Rather than seeing Theo's struggles as a personal failing, she reframes them as systemic challenges they can address together.

With this new understanding, Alice and Theo can work on real solutions—whether it's refactoring the codebase, introducing test automation, or hiring specialized team members to manage complexity. By cultivating a supportive environment where frustrations are met with empathy and problem-solving, Alice not only improves Theo's experience but also strengthens the organization's overall system. This approach creates a foundation for sustainable growth and success, transforming challenges into opportunities for alignment and progress.

The Consequences of Ignoring Systemic Complexity

When a CEO assumes that challenges stem solely from a single individual's failure—such as blaming the CTO for engineering inefficiencies—it sets off a destructive chain of events. Instead of addressing systemic issues, the CEO places blame on one person, creating a culture of fear and distrust. Often, this results in the CTO being pushed out, but the underlying problems remain. Without addressing the real sources of complexity, the same challenges will resurface with the next CTO, leaving the company stuck in a cycle of dysfunction.

This short-sighted approach weakens the organization, preventing it from achieving clarity and alignment. Instead of continually replacing individuals, leaders must develop the ability to diagnose systemic issues and implement structural solutions. By embracing empathy and systems thinking, organizations can break free from reactive decision-making and build a foundation for long-term success.

Building a Shared Understanding

When we recognize ourselves as integral parts of the complex system in which we operate, a significant shift occurs. Instead of working in silos or focusing solely on individual responsibilities, we begin to see how our actions ripple through the system, influencing its overall effectiveness.

This shared perspective fosters a sense of unity and common purpose, where everyone acknowledges their role in navigating and shaping the system. By embracing this interconnected view, we can approach challenges collaboratively, addressing systemic issues together rather than attempting isolated fixes that may only create further misalignment.

This collective awareness not only strengthens teamwork but also creates a more cohesive and resilient organization. When leaders and teams understand the dynamics of the system and their influence within it, they're better equipped to adapt to changes, overcome obstacles, and seize opportunities as a unified force.

This mindset allows the organization to function more fluidly, maintaining alignment even in the face of external pressures or internal complexity.

This collaborative approach doesn't just solve problems; it builds trust, enhances communication, and ensures everyone is aligned toward a common vision, making the entire organization stronger and more effective.

Liquid

Chapter 17:
Explore the State of the Systems Within Your Business

Speed is essential for maintaining agility and delivering results as your business grows. However, complexity can creep in, slow progress, and create bottlenecks.

These questions are designed to help you and your team uncover areas where complexity is impacting your speed and identify opportunities to streamline and improve:

- What recurring issues or bottlenecks are slowing down your delivery processes?

- Where has complexity crept into your workflows, and how can it be simplified?

- Are there key dependencies between teams or systems that are creating delays?

- How aligned are your teams on priorities, and where do misalignments occur?

- Have you noticed any shifts in how long it takes to deliver features or meet goals?

- What boundaries could you establish to give teams more autonomy and reduce overlap?

- Where can automation or better tools reduce manual effort and improve flow?

By addressing these questions, you can pinpoint areas where improvements are needed and take meaningful steps to ensure Speed remains a strength in your business.

PART 3:

Understanding the System

I don't know if you're ready to see what I want to show you, but unfortunately, you and I have run out of time. They're coming for you, Neo, and I don't know what they're going to do.

–Morpheus, *The Matrix*

Liquid

Chapter 18:

Introducing the Four Sentinels

As you may recall from Part 1, we mentioned the CTO Sentinel framework as a way to truly understand and manage the intricate dynamics within our organizations. There are four key sentinels that are indicators of the effectiveness of the complex system operating within a technology organization: Speed, Stretch, Shield, and Sales.

These four sentinels will help you identify when your system is flowing smoothly, when it's approaching a crisis, and when corrective action is needed.

Each sentinel represents a key aspect of how we influence the system at work in our businesses, and the CTO Sentinel successfully emerges as a vital property of a well-balanced organization that evolves and adapts as the business grows:

- **Speed** focuses on the ability to move quickly, whether shipping features, making decisions, or adapting to market changes.

- **Stretch** refers to the ability of the people and the organization to adapt and scale while maintaining alignment and preventing burnout.

- **Shield** relates to protecting the business by ensuring security, data, and regulatory compliance are effectively managed without becoming burdensome.

- **Sales** encompasses both effective external sales and the internal selling of ideas across the company, ensuring alignment on priorities and initiatives.

In the previous section, we delved deeply into **Speed**, where Alice and Theo encountered the challenges of increasing complexity and the need for better systems to maintain productivity.

Now, we'll begin to introduce the remaining sentinels, starting with **Shield** and **Stretch**, as they highlight the next layers of complexity that businesses encounter as they grow. Shield and Stretch will help you recognize when your system is being overstretched and when it needs protection, allowing you to adapt before the system begins to either boil or freeze.

Understanding these sentinels gives you a framework for identifying and managing the flow of complexity within your business, enabling you to lead with greater insight and foresight as your company evolves.

Speed as an Emergent Property

As we explored in Part 2, growing our businesses often starts with the need for Speed. To capture growth within our business, we need to move quickly and efficiently; however, the growing complexity of the underlying hidden system operating within our businesses creates a drag that slows our progress down.

Our goal is to cultivate a company that can adapt quickly and maintain momentum while navigating the inevitable complexities of growth.

Achieving this requires a thoughtful approach that goes beyond simply working faster or adding more resources. It means designing systems and clear boundaries within our business that encourage adaptability, allowing us to address the root causes of complexity rather than continuing to simply react to its symptoms.

Speed isn't just about how fast we can deliver products or services to an ever-changing market. Speed emerges when we build the sustained skills of agility, aligned internal teams, solid technical foundations, and clear decision-making processes. Investing in our full software management life cycle is key to Speed emerging as a property we can leverage and count on. By being intentional about the factors that lead to Speed, we can build a resilient foundation for sustainable growth, ensuring our company thrives both now and in the future.

> Speed emerges as we remove the friction within the system.

We saw this in the challenges Alice and Theo faced, where their system began to "boil" under the pressure of rapid growth and increasing complexity. It wasn't until they determined a course of action that redefined the boundaries within the system by expanding their team that they eventually found more stability and became productive again. The desire for growth is just one aspect that puts pressure on the complex adaptive system (CAS) of your business to deliver sustained speed.

The Pillars of Growth and Protection

In the world of CAS, **Stretch** and **Shield** are fundamental forces that guide an organization's journey through growth and complexity. These two sentinels are not opposites but partners, working together to ensure a business can adapt, evolve, and safeguard its operations in an unpredictable environment.

Stretch is the organization's (and its people's) capacity to push its limits and embrace the discomfort of growth. It's the energy that drives innovation and evolution, compelling teams to challenge the status quo and explore new possibilities. Stretch is evident when a company pursues bold goals or adapts to unforeseen challenges with creativity and agility. It thrives in environments where risk-taking is encouraged and where the desire to improve is a collective mindset. Stretch isn't about reckless expansion but about creating conditions where the organization can adapt and thrive in the face of complexity.

On the other hand, **Shield** represents the protective systems that preserve an organization's resilience and sustainability. It's the force that mitigates risks, ensures stability, and safeguards the company's resources and capabilities. Shield is the mechanism that prevents chaos from overtaking innovation. Whether through robust infrastructure, clear processes, or thoughtful risk management, Shield creates the boundaries that allow the organization to stretch without breaking. It defends against external threats, such as market disruptions or compliance failures, and internal risks, such as employee turnover or unmanageable technical debt.

The interplay between Stretch and Shield is dynamic and essential. Too much Stretch without the balancing force of Shield can push an organization into chaos, where the rush to innovate outpaces the systems needed to sustain progress (boiling). Conversely, an overemphasis on Shield without Stretch can lead to stagnation, as excessive caution stifles creativity and prevents the organization from adapting to change (frozen).

The balance of these two forces is an indicator of a thriving system. *Stretch fuels growth, while Shield ensures that growth is sustainable.* Together, they allow an organization to operate in the liquid phase of complexity, where adaptability and resilience coexist.

In the chapters that follow, we'll explore how these sentinels illustrate correlations within your organization, helping you to see and understand the patterns that drive your business forward.

Liquid

Chapter 19:
Understanding Correlations in a Complex Adaptive System

In the context of a business, understanding correlations within our complex adaptive system (CAS) is critical for identifying whether subsystems are operating in a **frozen, boiling,** or **liquid** state.

Correlation refers to the degree of interdependence or connection between elements within the system. In a business, this describes how actions, decisions, or outcomes in one area influence or are influenced by those in another.

Strong correlations signify tightly linked elements, where changes in one part have significant ripple effects across the system. Weak correlations indicate less interdependence, often leading to fragmentation or inefficiency. Recognizing and managing these relationships is essential to navigate complexity, identify patterns, and steer the organization toward sustainable growth.

These correlations—the connections and interdependencies between different parts of the organization—serve as the framework through which resources, decisions, and information flow, shaping the system's overall effectiveness. Much like the interactions between water molecules in solid, liquid, and gaseous states, the correlations within a company vary depending on the system's state.

A **frozen** system is overly rigid, with tight correlations that inhibit flexibility and stifle innovation. A **boiling** system lacks cohesion; correlations have broken down entirely, creating chaos and disarray. A **liquid** system strikes the ideal balance, with correlations that are dynamic yet cohesive, enabling the organization to adapt effectively while maintaining alignment and purpose.

Correlations Through the Lens of Stretch and Shield

The dynamics of Stretch and Shield are deeply intertwined with these states. Stretch thrives on the freedom and adaptability of a liquid state, where teams can innovate, experiment, and evolve without being hindered by excessive rigidity or lack of coordination. Shield, on the other hand, ensures that, even in a liquid state, the organization is protected from instability, providing the structure and boundaries needed to sustain progress without descending into chaos.

In a **frozen** state, correlations are excessively tight, creating barriers to Stretch. Teams may feel constrained by inflexible processes or dependencies, unable to explore new ideas or respond to change. Shield dominates excessively in this state, leading to overregulation and an environment where innovation is stifled.

In a **boiling** state, correlations have weakened or broken down entirely, leaving parts of the organization disconnected and vulnerable. Stretch becomes unfocused as teams work at cross-purposes or pursue misaligned goals. Shield is almost absent, leaving the organization exposed to risks and inefficiencies that threaten its stability.

A **liquid** state represents the ideal equilibrium. Correlations in this state are neither overly rigid nor entirely disconnected. Teams have the flexibility to stretch and innovate, while Shield ensures that interactions remain aligned, protecting the organization from destabilizing forces. This balance fosters a system where adaptability and resilience coexist.

Influencing System Dynamics

Correlations are the connective tissue of the organization, revealing how one part influences or is influenced by another. By actively assessing these relationships, we can identify points of rigidity, disconnection, or balance and apply the right interventions to optimize the system's effectiveness.

In a **frozen** state, where correlations are overly rigid, the organization is trapped by dependencies, approvals, and processes that restrict adaptability and progress. Leaders may sense this as a sluggishness in delivery or an inability to pivot quickly in response to market demands.

To navigate out of this frozen state, leaders must actively assess their teams and processes for points of rigidity. This starts with asking: Where do approvals slow us down?

Which handoffs between teams repeatedly cause delays? Are there areas where teams hesitate to act without explicit permission? Once these constraints are identified, loosening them requires intentional action. This might involve streamlining multistep approval chains, delegating more decision-making power to teams, or revisiting legacy policies that no longer serve the company's current needs.

Creating space for Stretch to emerge means reducing friction—empowering teams with autonomy while retaining clarity around goals. Leaders who shift from gatekeepers to enablers can unlock the potential trapped within a frozen system.

In a **boiling** state, where correlations have broken down, the organization is marked by chaos, misalignment, and unpredictable delivery. Leaders often notice this state when teams operate in silos, duplicate work emerges, or decisions stall due to confusion over ownership. To assess this state, leaders can ask: Where are teams working independently but producing conflicting outcomes? Where do decisions frequently escalate because roles are unclear?

Calming a boiling system requires adding stabilizing structure without stifling progress. This can involve introducing cross-team planning rhythms, clarifying who owns key decisions, or formalizing agreements on how priorities can shift. By strengthening these connections, leaders can restore Shield and guide the system back into a coordinated, productive flow.

Encouraging flow within a **liquid** state requires consistent monitoring and fine-tuning. By leveraging insights from the system's correlations, we can implement changes that nurture adaptability without sacrificing stability. This involves recognizing when to push boundaries and innovate through stretching and when to ensure the system remains resilient and sustainable by strengthening our ability to shield.

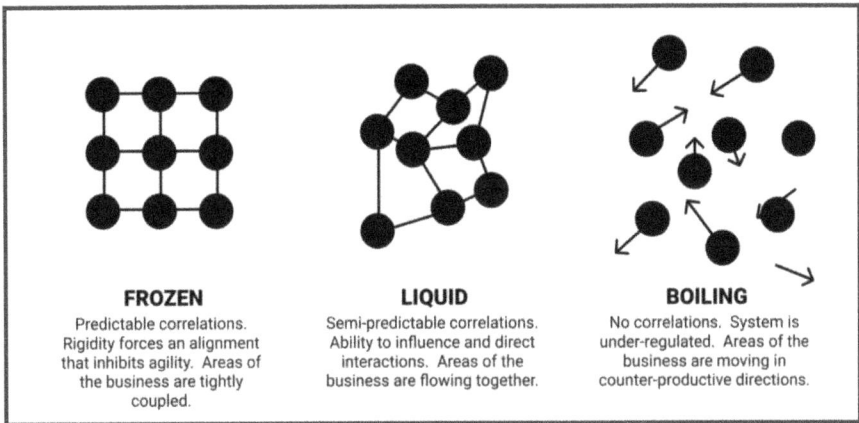

FROZEN	LIQUID	BOILING
Predictable correlations. Rigidity forces an alignment that inhibits agility. Areas of the business are tightly coupled.	Semi-predictable correlations. Ability to influence and direct interactions. Areas of the business are flowing together.	No correlations. System is under-regulated. Areas of the business are moving in counter-productive directions.

Understanding and managing correlations enables us to keep the organization in a productive, adaptable liquid state.

By continuously evaluating how Stretch and Shield interact with these correlations, leaders can cultivate an environment where flexibility and alignment coexist, driving innovation, protecting the system, and ensuring long-term success. Recognizing these patterns and their impact is a critical step in mastering the dynamics of CAS and guiding your organization toward equilibrium and growth.

Liquid

Chapter 20:

Sentinels Under Pressure

When systems reach their extremes, whether frozen in rigidity or boiling in chaos, the dynamics of Stretch and Shield are tested. These states push the limits of an organization's ability to adapt and protect itself, often revealing critical weaknesses or imbalances.

In this chapter, we explore how Stretch and Shield manifest under pressure, examining their roles in both frozen and boiling systems. Understanding these dynamics provides valuable insights into recognizing and resolving these extreme states, guiding the organization back to balance and flow.

Frozen Systems: Overregulated

Just as ice forms when water molecules become tightly correlated, so too does an organization freeze when individual actions are overly dependent on each other.

When every team member is dependent on one individual to complete their work, the entire system becomes bottlenecked and overly constrained by that person's output. This overcorrelation stifles progress and creates inefficiencies, as the team's momentum is entirely dictated by the availability and productivity of a single point of dependency. A common example of this occurs in technical teams with separate front-end and back-end developers, where progress on one side is stalled while waiting for the other to complete critical tasks. Similarly, a team that relies on a single key individual to review, approve, or execute all work finds itself hampered by overcorrelation. These situations leave the system rigid and vulnerable, unable to adapt quickly or operate at its full potential.

When too many roles and tasks are dependent on each other, individual flexibility is lost. This can cause bottlenecks, reduce innovation, and lead to frustration, as everyone is locked into a specific order of operations and blocking each other.

The leader who doesn't understand this might call for more meetings, require more synchronized efforts between team members, and perhaps tighten feedback loops to the point of micromanagement. We might not blame this leader for trying harder to coordinate efforts between team members. But we know better than to address the symptoms rather than the underlying state of the system.

Leaders who understand the dynamics of frozen systems play a crucial role in restoring balance, where Stretch and Shield can coexist harmoniously. Recognizing that rigidity is a symptom of excessive correlations, they can take action to reduce dependencies by creating clearer team boundaries, establishing more independent roles, and delegating authority effectively.

When they introduce adaptability into overly rigid processes, leaders unlock the potential for growth and innovation, which preserves the stability needed to sustain progress. In this dynamic equilibrium, teams are empowered to adapt, innovate, and drive the organization forward while remaining protected against unnecessary risks.

In a frozen system, correlations between elements are excessively tight, creating a rigid environment where every action is overly dependent on others. This rigidity often stems from a well-intentioned but overactive Shield sentinel. Processes and structures designed to protect the organization, such as compliance measures, approval workflows, or hierarchical decision-making, become so inflexible that they stifle innovation and adaptability. While Shield aims to provide stability, it overcorrects in frozen systems, leaving little room for the Stretch sentinel to thrive.

When Stretch is suppressed in this way, teams struggle to take risks, adapt to changing circumstances, or innovate effectively. Creativity is dampened by an overwhelming focus on control, leading to stagnation.

For example, in a product team, a frozen state might require every feature or code update to pass through multiple layers of approval, delaying deployment and frustrating team members who are eager to respond to market demands.

Imagine a business where a marketing campaign is entirely dependent on the completion of a product update, but the product team can't make progress because they're waiting on approvals from legal. This creates a cascading effect: If one team is delayed, the entire organization stalls. Much like a substance in a frozen physical state, where knowing the position of one molecule allows us to predict the position of others, a frozen system in business means that every action is constrained by interdependencies, leaving the company unable to move forward effectively. The cultural impact of frozen systems is significant.

Employees often feel undervalued, constrained, and disconnected from the company's goals. The lack of flexibility discourages engagement and creativity, contributing to frustration and burnout. Over time, this environment erodes trust and morale, resulting in high turnover rates and further entrenching the organization in its rigidity.

To escape a frozen state, leaders must recalibrate the balance between Stretch and Shield. This involves loosening excessive Shield measures while preserving essential protective functions. To resolve these interdependencies, the company must first identify and untangle the critical bottlenecks causing the freeze. This involves creating **clearer boundaries** between teams to reduce unnecessary dependencies and empowering each team to move forward independently within defined scopes. Legal approvals could be streamlined by predefining compliance templates or thresholds for low-risk changes, allowing the product team to progress without constant oversight. Marketing can align on a more flexible campaign strategy that accounts for potential shifts in product timelines, mitigating the downstream impact of delays.

In addition to boundaries, leadership must prioritize cross-functional alignment by improving **communication and collaboration frameworks**. Introducing tools like shared project trackers or synchronized sprint planning sessions enables teams to better anticipate interdependencies and resolve conflicts earlier in the process. Teams that operate with a degree of autonomy yet remain loosely coordinated are more resilient, as progress can continue even if one part of the system encounters delays.

Fostering a culture of experimentation can encourage Stretch by enabling teams to test and iterate new ideas within well-defined boundaries.

This approach allows the company to shift from a frozen state to a more flexible, liquid state, where interdependencies are managed thoughtfully and progress can flow freely across the organization.

Boiling Systems: Underregulated

Boiling systems emerge when the connections between elements in a business become too loose, creating an environment of chaos and disorganization. In this state, correlations break down entirely, leaving teams and departments uncoordinated and operating independently without alignment to shared goals or strategies. This lack of cohesion undermines the organization's ability to function effectively, resulting in inefficiencies, missed opportunities, and frustration.

The Stretch sentinel often dominates in a boiling system, but its influence becomes scattered and unfocused. Teams may take bold risks and attempt to innovate, but without Shield's balancing structure, these efforts often lead to misaligned priorities and wasted energy.

A sales team might independently promise features or timelines to clients without consulting product or engineering teams, creating a disconnect between what is sold and what can realistically be delivered. This misalignment ripples through the organization, leading to rushed development, missed deadlines, and strained relationships both internally and with customers.

On the other hand, the Shield sentinel is underactive in boiling systems, leaving the organization vulnerable to errors, inefficiencies, and reputational risks. Without clear processes or boundaries, teams struggle to coordinate their efforts and critical protective measures, like quality assurance or compliance checks, may be overlooked. The absence of Shield allows the system to spiral further into chaos, with no mechanisms in place to bring it back into alignment.

Leadership in a boiling system often feels reactive rather than proactive, as disorganization compels leaders to focus on firefighting immediate issues instead of driving long-term strategy. The lack of clear accountability and structure exacerbates this challenge, enabling individuals to "do their own thing" without alignment to organizational priorities.

When people are given excessive autonomy without clear boundaries or defined roles, they may stray from their core responsibilities, pursuing personal interests or priorities at the expense of team and organizational goals. This dynamic creates confusion, duplication of efforts, and uncoordinated decision-making, further eroding efficiency.

The leader who doesn't understand that they're in a boiling system may start singling out members of the engineering team as not being team players, with an eye on letting them go. They may also fall into the trap of placating their engineers, buying their allegiance and focus through bonuses and bloated compensation packages. Leaders, overwhelmed by the chaos of a boiling system, often resort to overcorrection or symptom mitigation.

These reactive measures frequently tip the system into a frozen state, stifling innovation and adaptability instead of restoring balance. To effectively address a boiling system, leaders must establish accountability frameworks that align individual autonomy with clear expectations and shared objectives, ensuring both flexibility and cohesion. They need to recalibrate the relationship between Stretch and Shield, ensuring innovation is channeled productively while reintroducing necessary structure.

Leaders who think in systems, however, can address boiling systems by implementing common frameworks, such as shared documents, communication protocols, or standard operating procedures, that provide a baseline for coordination and alignment. These tools act as guideposts, ensuring that while teams and individuals retain their independence, their efforts remain aligned with the organization's broader goals.

For example, a collaboratively developed strategic plan that outlines key business objectives, product priorities, and delivery timelines helps sales, product, and engineering teams align on what will be built and when. This shared plan provides clarity on upcoming features, expected delivery dates, and trade-offs, enabling teams to make realistic commitments to customers while avoiding conflicting priorities. Regular checkpoints to review and adjust the plan ensure that all departments stay in sync as market conditions and business needs evolve. By establishing these frameworks, leaders create a system where teams are empowered to make decisions within their boundaries but remain connected to the organization's strategy.

This balance prevents the chaos of a boiling system while avoiding the rigidity of overcorrelation. Over time, these shared structures build trust and predictability, enabling the organization to function more cohesively and respond more effectively to both challenges and opportunities.

With the right frameworks in place, leaders can guide their organizations toward a productive, adaptable liquid state, where creativity and coordination coexist in harmony.

The Liquid State: Well-Regulated Balance

The liquid state is akin to the properties of water—dynamic, adaptable, and fluid. Just as water flows smoothly and adjusts to its environment, a business in the liquid state maintains flexibility while staying cohesive. This state represents a well-regulated balance, where teams operate with both the freedom to innovate and the structure to remain resilient.

It's within this state that the interplay between the Stretch and Shield sentinels is most harmonious, creating an environment where adaptability and stability coexist.

In the liquid state, Stretch thrives. Teams are empowered to experiment, take calculated risks, and innovate, knowing they have the autonomy to adapt quickly to changing circumstances. However, Stretch operates most effectively when anchored by Shield, which provides the protective structures needed to safeguard against instability and chaos. Shield establishes the boundaries within which experimentation can occur safely, ensuring that risks are managed and disruptions are minimized. For example, a product team in a liquid state might explore bold new features or enhancements, leveraging Stretch to respond creatively to customer needs. At the same time, Shield ensures these innovations are tested thoroughly and deployed within reliable processes, preserving the overall integrity of the system. This balance allows the organization to move quickly without sacrificing quality or long-term stability.

Leadership plays a pivotal role in sustaining the liquid state. Leaders act as mediators between Stretch and Shield, fine-tuning the balance to ensure neither sentinel dominates. They create environments where creativity and structure coexist by encouraging cross-functional alignment, maintaining clear but flexible processes, and promoting open communication. In this state, goals are shared across teams and collaboration flows naturally, reducing friction and amplifying impact.

While the liquid state is the ideal, it's not without its imperfections. The dynamic interplay between Stretch and Shield creates natural tensions that must be well-managed. Rapid innovation may occasionally pressure stability, while protective measures might sometimes slow experimentation.

These tensions aren't signs of failure but indicators of a healthy, evolving system that requires ongoing calibration. To maintain the liquid state, leaders must continuously assess and adjust the system's dynamics.

This includes evaluating whether teams have the autonomy they need to innovate (Stretch) and ensuring that processes provide sufficient structure to protect against risks (Shield).

Regular retrospectives, alignment sessions, and process reviews can help identify areas where recalibration is needed, ensuring the organization remains adaptable and resilient.

The liquid state offers a unique advantage: It cultivates an environment where teams can adapt, innovate, and collaborate effectively while remaining protected against unnecessary risks.

A liquid state doesn't mean everything is perfect. It means the system is designed to adapt, grow, and remain resilient as conditions change, allowing the business to maintain both momentum and stability. It's the ideal operating mode for businesses seeking to manage complexity effectively while driving innovation and sustaining productivity.

How difficult is it for a growing company to remain liquid? Let's see how Alice and Theo are faring as their company continues to grow and their business becomes more and more complex. How do they adapt when parts of the business start to boil and freeze?

Chapter 21:

Alice's Company Is Growing Fast

Alice and Theo's business is growing fast. Their B2C product is performing well; however, closing consumer business a little at a time is preventing them from being truly profitable. Users love the product, but consumer growth starts to substantially slow down.

Alice is approached by a large enterprise to launch a B2B version for the enterprise's customer base. While the features are largely the same as the B2C product, it's for a closed community, which will have a different pricing and support model.

Neither Alice nor Theo realizes that they're now entering a new phase of complexity—one focused on *shielding* their business from potential threats as they expand.

Mistake #1: Building a Monoteam

Alice walks into the office, filled with excitement about the new opportunity the enterprise client had presented. "Theo, this is huge! We have a chance to build out a B2B version of our product. We just need to scale our team, build faster, and get these features out the door."

Theo looks up from his laptop, fear in his eyes at the many roadblocks they'll need to overcome to be able to deliver what Alice is asking. "I get it, Alice. But we're already juggling a lot with the B2C product. Taking on more features, especially for enterprise clients, means tighter security, more compliance checks, and new architecture."

Alice, undeterred, smiles. "As you suggested before, we can solve this by hiring more developers. We've been working with just a few engineers. If we double the team, we can knock out both B2C and B2B features at the same time."

Theo hesitates: "I don't know, Alice. Adding more people isn't always the solution. If we keep them all working together on the same projects, the team might get bogged down trying to coordinate."

But Alice is set on her plan. "I understand your concerns, Theo, but we need speed. We can handle it. Just think: If we hire more people, we'll be able to split the work, right?"

"Yeah, theoretically," Theo says, unconvinced but willing to give it a try. "I'll integrate the new hires into the team, and we'll start getting everyone on the same page for both B2C and B2B development."

The Monoteam Is Boiling

In the following weeks, Alice and Theo ramp up hiring, bringing in new developers. Theo, true to his word, integrates them into the existing team. The developers quickly start working on features for both the B2C and B2B products, handling everything from UI improvements to back-end security protocols.

At first, it seems like things are moving faster. But soon, coordination issues begin to crop up. Theo finds himself constantly in meetings, aligning the new team members with the veterans, explaining dependencies, and resolving conflicts over priorities.

Every time a new feature is added, it introduces more complexity into the codebase. One afternoon, Theo drops by Alice's office. "We've got a problem," he says, visibly stressed. "The more people we add to this team, the slower things are getting. Everyone's working on both B2C and B2B features simultaneously, and it's becoming impossible to keep everything aligned and coordinated. Every decision now takes twice as long and every developer has to know everything about our entire system to accomplish their tasks."

Alice frowns. "But we hired more people to increase velocity, not slow it down."

"I know, but more people aren't reducing the complexity; they're adding to it," Theo explains. "The team is spending more time coordinating than building. The more developers we add, the more chaotic the projects feel. If we keep pushing like this, we're going to hit a breaking point. Perhaps we already have and are just now realizing it."

Realizing the Systemic Issue

Alice leans back, considering Theo's words, but the anxiety tightens in her chest. They promised the B2B client progress, and that launch timeline is looming. If they fail, it could damage their reputation before they've even gotten started. She exhales sharply. "So, what do we do? We need to deliver both the B2C and B2B products. Other companies seem to grow their teams and gain speed. Why is this not working for us? What are we getting wrong?"

Theo feels his pulse quicken. He's been asking himself the same thing. Every day, it feels like he's fighting to keep the machine from falling apart. The more they build, the slower everything gets. "We need to separate the work," he says, trying to sound more confident than he feels. "We can't have everyone working on everything.

It's like the team is tangled up. We need to split into smaller teams so each group can focus."

Alice nods slowly, the reality of the situation sinking in. She sees Theo's exhaustion, but also his determination. Her mind begins to map out the implications: restructuring the team, setting clear boundaries, communicating this shift without eroding morale.

"You're right," she says, though the words feel heavier than she expected. "If we don't help people focus, we're just going to keep shoveling features onto the fire until everything grinds to a halt."

But even as she agrees, doubt lingers. Will the team accept this shift? Will they see it as a sign of leadership or of panic? What if this is just the beginning? What other cracks are forming beneath the surface?

They sit in silence for a moment, both feeling the weight of what's ahead. This isn't just about a team restructure. It's the first real test of whether their startup can survive scaling.

The Boiling System: Pushing Toward the Edge

Alice and Theo are no longer guiding a fast-growing startup. They're fighting to keep it from coming apart. Growth has given way to chaos. The company is no longer flowing along; it's boiling over.

The monoteam structure, with developers working interchangeably on both B2C and B2B products, has resulted in an overburdened system where coordination is becoming more complex by the day. **Brooks' law,** coined by Fredrick Brooks in his book *The Mythical Man-Month,*[14] observes that *adding people to a late software project will make it later.*

Why is this the case? When people are added to a project, they have ramp-up time. Existing team members have to be pulled off the line to help the new team members understand the system sufficiently to be productive, thus also delaying any work that's currently in flight by the existing team members.

The overhead of communication across the people on the team grows. This is called *combinatorial explosion*. As every person needs to stay in sync with each other, the team quickly finds themselves in meetings every day attempting to stay in sync, but not actually doing the required work to meet the goal.

Software development isn't the same as easily divisible work tasks, which are much easier to scale. Software development requires deep understanding of the existing system and codebase, and often requires specialized skill sets to complete the mission.

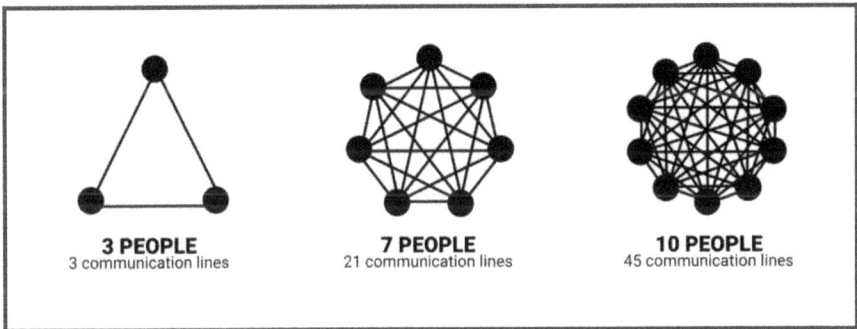

3 PEOPLE
3 communication lines

7 PEOPLE
21 communication lines

10 PEOPLE
45 communication lines

Brooks' law of increasing communication complexity

(The Mythical Man Month, 1975)

If growth isn't well-managed, with efficient lines of communication maintained, the overhead of managing these lines of communication often outweighs the benefits of additional hands on deck.

The team will become frozen as a result of the complexity of the communication. It's not possible for everyone to work on everything simultaneously and always be "in the know" on it all.

As Alice and Theo add more developers without creating clear boundaries, they inadvertently increase the correlations between tasks and roles, making it difficult for anyone to work independently. Developers spend more time synchronizing their efforts, coordinating on even small decisions, and untangling overlapping responsibilities rather than focusing on the features they need to deliver.

In a boiling system, where complexity is high and increasing, delays and miscommunications multiply. Each feature that gets added introduces more dependencies, and the system begins to lose its flexibility. Every action taken by one part of the team creates ripple effects that slow down the entire organization. Decision-making grinds to a halt as more people are brought in, yet productivity drops because the team is overwhelmed by the sheer amount of information they need to process.

If this situation isn't corrected, it won't be long before the team shifts from boiling to freezing (bypassing the desirable liquid state). When complexity grows unchecked, the system will become rigid. Teams will have no choice but to establish informal workarounds, eventually leading to paralysis. When a system freezes, it's no longer adaptable.

Every person is waiting on another person, and every team is waiting on another team. Any change or new feature will take exponentially longer to develop and implement.

At this stage, progress will become nearly impossible, with developers locked into processes that stifle innovation and responsiveness.

The boiling point is a warning. Without decisive action, Theo's engineering team will inevitably freeze, locking them into a cycle of decreasing agility and increasing frustration. If the situation continues unchecked, the entire company will eventually freeze.

The solution lies in recognizing the signs early, when the system first starts to simmer, and taking steps to reduce the complexity before it solidifies into an immovable obstacle. By splitting teams, creating clear boundaries, and distributing responsibilities effectively, they can turn their boiling system back into a liquid state, where adaptability and flow are restored.

Beyond the growing complexity, Alice and Theo face a critical issue tied directly to the Shield sentinel: the risk of reputational damage if they fail to deliver on time for the B2B client. As they navigate the development of their new enterprise offering, the stakes are much higher than was the case for their previous consumer-facing product.

Delays, missed deadlines, or inadequate features could not only result in them losing this major client but also damage the company's reputation in the broader market. B2B clients expect reliability and the delivery of promised features on schedule, and any misstep in meeting those expectations can ripple outward, affecting future business prospects and trust within the industry. By continuing to operate in a boiling system with blurred boundaries, they risk the delivery of the B2B product slipping further behind, potentially eroding the trust their new client has placed in them. To shield the company from these reputational risks, Alice and Theo need to take swift action to reduce complexity, split the teams, and ensure both products receive the focused attention they need. Prioritizing the integrity of their delivery timelines and quality will help them maintain their standing in the market and grow sustainably.

Alice and Theo are faced with *organizational complexity*. They must get this under control or their reputation as a business will be at risk.

The Turning Point

Together, Alice and Theo devise a plan to reorganize the monoteam. Theo suggests dividing the developers into two teams: one focused on the core B2C product and another on the features specific to the new B2B product. By creating clear boundaries between the teams and better aligning their work to the business, they hope to reduce the need for constant coordination and make both teams more efficient.

Theo explains, "With this split, each team will have a clearer focus and set of responsibilities. Right now, things are getting tangled because everyone's working on everything, and it's creating confusion and slowdowns. By dividing the work and setting clearer boundaries, we can help everyone stay on track and work more efficiently. It'll make things feel less chaotic and more manageable."

After splitting the teams, Alice and Theo agree that the B2B and B2C teams will focus on the distinct needs of their respective domains. The B2B team takes on enterprise-specific features, such as managing sensitive data, meeting regulatory compliance requirements, enabling single sign-on, and building a robust multi-tenant architecture to securely serve multiple clients within a single system.

Meanwhile, the B2C team concentrates on enhancing the core user experience for their existing customer base by improving usability, refining consumer-focused features, and driving growth through targeted updates.

To streamline collaboration, they decide that most cross-team interactions will happen at the team lead level. Individual contributors stay focused on identifying risks within their own teams, while team leads handle communication and resolve issues that span across groups.

Theo restructures the engineering organization, splitting the original complex adaptive system (CAS) into two interrelated subsystems. This change not only simplifies the complexity within each team but also creates a more focused and adaptable structure, which is better suited to the distinct demands of both the B2B and B2C products.

To support capabilities that are shared across both products—such as core infrastructure, payment processing, and customer support systems—they also form a dedicated shared services team. This team works alongside both product teams to ensure these foundational systems meet the needs of each product without becoming a bottleneck.

With these new team boundaries in place, Alice and Theo give each team the freedom to move forward independently while ensuring critical touchpoints remain aligned.

The teams can now progress with clarity and stability, avoiding the entanglement that had slowed them down before. In the weeks that follow, the organizational restructuring brings relief. The B2C team can now focus entirely on their product, while the B2B team works on the new features their client needs for launch.

During this turning point, Theo accidentally applied a key successful CAS *influence pattern* designed to address organizational complexity: He created new boundaries that aligned engineering directly to the flow of the business. With this move, the complexity in the system decreased and productivity increased. This move shields the company from reputational risk.

Neither Theo nor Alice realizes that their success is also partly due to the fact that the original team had already absorbed the company's culture, allowing them to adapt seamlessly to the new team structure.

They had dodged the immediate crisis, but the next challenge was looming: Theo's next hire would reveal just how important culture and mentorship are to maintain balance within these complex subsystems.

Theo thought he'd found the perfect solution and reasoned that creating smaller teams was the way to go. However, he didn't realize that his success was partly accidental. The new team was made up of individuals who'd already absorbed the company's culture, which allowed for a smoother transition. This assumption led Theo to make another, more dangerous mistake.

Mistake #2: Theo's New Team of New People

After the success of the first team split, Theo decides to apply the same approach when they need to expand further. This time, however, Theo hires entirely new employees and forms a brand-new team to focus on the core billing, subscriptions, and entitlements work that will span both the B2C and B2B products.

What Theo doesn't anticipate is that this new team lacks the cultural understanding of the original team. They have yet to experience the company's deep commitment to delivering flawless demos or learn the nuances of how teams interact and share responsibilities. The symptoms of this disconnect show up quickly:

- *Ignorance of culture: The new hires don't take new feature demos seriously because they don't realize the original team views demos as a point of pride in their accomplishments. The veterans would stress about perfecting their demos, but the new team misses this cue and instead treats demos as a necessary evil.*

- *Poor code hygiene: The new team isn't familiar with the established coding habits and best practices that Theo's original team have developed over time. They lack the informal training that comes from working alongside experienced team members.*

- *Misunderstandings about project inclusion: The new team struggles with not being included in every decision. They don't realize that, culturally, it isn't necessary for them to weigh in on everything, and that the established team trusts the decisions made in their absence. This leads to confusion and frustration as the new hires try to involve themselves in every step of the process, slowing things down further.*

By creating a completely new team of outsiders, Theo has inadvertently replicated Alice's initial mistake of failing to recognize the importance of cultural and process alignment. He has used the wrong pattern for team creation, one that ignored the need for mentorship and culture transfer within the teams.

In their rush to speed up the organization's delivery, they failed to protect the organization from organizational complexity.

Theo sits at his desk, staring at the performance reports of the newly formed team. The numbers aren't adding up. Productivity is low, and the quality of their recent demo was well below the company's standard. It was easy when he split the existing team into two teams. Why wasn't it working when he added a new team? Why does everything still feel so hard?

He flips through his notes from the past few weeks. Small misunderstandings have ballooned into delays, design decisions need constant correction, and every demo review leaves him and Alice exchanging uneasy glances. The new team's work lacks the polish and pride that the original staff made second nature.

Theo exhales slowly as the pattern becomes clearer. The veterans operate on shared instincts—they know what quality looks like, when to escalate issues, and how to push to the finish line. None of this is written down; it lives in the culture, built through experience. The new team has none of that.

They're working in isolation, disconnected from the rhythms and expectations that fuel the rest of the company. Without experienced team members to guide them and help them normalize to the culture, they're left guessing and creating their own. And Theo is left constantly re-explaining goals, checking work, and filling in gaps. The structure looks right on paper, but the system is still boiling—because culture wasn't part of the design. When Alice stops by, he explains, "I think I messed up by building a completely new team from scratch.

They don't have anyone to show them how we work, how we care about things like demos and coding standards. It now feels like I'm managing two completely different companies and processes."

Alice nods in agreement. "Yeah, the last demo didn't seem like our usual work, and the business said it wasn't close enough to what they asked for. So, what's your plan?"

"I need to integrate some of the experienced team members into this group," Theo replies. "They'll help transfer the culture. It will help the new team better acclimate to how we work, what we value, and how we handle code, and will also role-model how we take pride in what we do."

Theo wastes no time. The next day, he gathers the new team and explains the changes. He rotates experienced developers into their group to mentor and guide them, ensuring the culture and work standards are passed down.

With this simple adjustment, Theo corrects the alignment gap and quickly sees improvement. The new hires begin to understand the company's values, and the team's performance starts to reflect the standards that made Alice and Theo's company successful in the first place.

Proactive Evolution

After resolving the complexity caused by the monoteam strategy and successfully restoring their system to a liquid state, Alice and Theo are able to bring more focus to their delivery and produce accelerated results. Theo's time is freed up to proactively consider future threats.

By both reducing organizational complexity and influencing culture, the company has accelerated the delivery of their B2B product. Alice and Theo understand, however, that moving forward without a proper security and governance program would expose them to significant risks. They know that in the enterprise space, data security and regulatory compliance are paramount. Alice and Theo take a moment to assess their approach. Rather than pushing the system harder, they recognize the need to further shield the organization for sustainable growth.

To do this, they hire a cybersecurity expert to oversee the implementation of enterprise-level security protocols and a compliance officer to ensure they meet industry standards for data privacy and regulatory requirements. These new hires take on the critical work of protecting the organization from potential security breaches, privacy violations, and noncompliance penalties, effectively lightening Theo's load.

The company soon begins conducting regular audits of their internal processes to ensure cross-team collaboration and knowledge transfer remain streamlined. By doing this, they fortify their culture while creating a structured, scalable system that can support both the B2C and B2B products without overwhelming the teams. Establishing these boundaries ensures that as new people join, they onboard quickly and adapt to the company's standards without complicating the workflow.

This proactive approach allows Alice and Theo to move confidently into the next phase of growth.

They acted early, when the problem was just starting to simmer, which helped them adjust quickly to remain in the liquid state.

With security and compliance now woven into the fabric of the company, they have successfully shielded their organization from potential external risks, all while maintaining a flexible, adaptable system that flows efficiently.

Liquid

Chapter 22:

Growing Up Successfully

When growing a company, having a strong foundational structure is just as important as achieving the desired outcome. This structure determines how teams, processes, and strategies are built and interact, ensuring that growth is intentional and sustainable.

To truly shield our company from these recurring threats, we need more than reactive fixes. We need an underlying blueprint that supports growth while keeping complexity in check. This is where the concept of an ontological pattern comes in.

This is a deliberate approach to designing the foundational elements that guide how an organization evolves. By embedding this pattern into our structure, we can better anticipate challenges, seize opportunities, and build a system capable of sustaining success as complexity increases.

What Is an Ontological Pattern?

A *pattern*, in this context, refers to the underlying framework or blueprint that dictates how a system evolves. It's not just about the final form but how individual components are designed to interact and develop over time. An *ontological pattern* goes deeper.

It's the core logic and principles that shape the very existence of the system, much like the DNA of a living organism. For example, in a company, an ontological pattern dictates not just the organizational chart but also how teams form, how culture is passed down, and how processes are organically shaped over time by the people involved.

It's a growth-first approach, focusing on how smaller systems scale into larger ones without losing coherence or adaptability. Instead of forcing structure on to people, you allow the structure to emerge as the people grow within the system, ensuring that the organization remains fluid and responsive as it evolves. Just like a seed contains the full genetic blueprint for a tree, an ontological pattern carries the essential framework for sustainable growth and adaptation.

Gall's law: Complex systems evolve from simpler ones over time.

In the 1977 book *Systemantics*, John Gall[15] asserts that complex systems evolve from simpler ones over time. And the only complex systems that actually work are those that evolved from smaller systems that worked. He also asserts the opposite, in that a system intentionally designed to be complex from the start will fail.

Imagine trying to instantly create a perfectly organized 50-person company by hiring 50 people and assigning them directly into predefined roles. The structure may eventually settle, but the chaos of that approach will lead to an operational disaster for a very long time. There's a massive difference between simply hiring people and placing them into a "perfect" structure versus organically growing a team from a smaller, manageable base.

Hiring five people and allowing them to grow naturally into their roles creates a system that can adapt, evolve, and, most importantly, develop its own processes and culture in a more stable path for growth.

This idea is central to the CTO Levels framework, especially at the lower levels (Levels 0–3), where the focus is on understanding what needs to be done and how to build strong foundational systems that can eventually scale.

You can't just impose solutions onto lower-level problems; you need to grow the solution. Think of it like farming. The farmer doesn't just make the corn; they grow it. The farmer nurtures the seed, and through a natural ontological process, over time the corn grows.

Contrast that with the hypothetical *Star Trek*[16] replicator, a machine from science fiction capable of assembling a person, molecule by molecule, on demand. The replicator represents a vision of perfect control and precision—every atom placed exactly where it belongs to create the desired outcome instantly. Yet the complexity required to design and operate such a machine is astronomical. It demands mastery over every microscopic interaction, with no room for error.

In reality, nature has already solved this problem through a far simpler, more resilient system: a seed. Plant it in good soil, give it sunlight and water, and it will grow—guided by patterns shaped over millennia. The process isn't instant, nor is it perfectly predictable, but it works, reliably and sustainably. So, we must ask ourselves: In our efforts to scale and systematize our companies, are we trying to build a *Star Trek* replicator—obsessing over control and complexity—when we should be planting seeds, nurturing growth, and letting robust patterns emerge naturally? In *Team Topologies*, Matthew Skelton and Manuel Pais[17] offer valuable guidance for structuring teams and aligning responsibilities, providing a strong foundation for organizational design. When refactoring an organization or building a new department, however, success requires more than simply adopting a predefined structure.

Leaders must also create an evolving ontological pattern that enables teams to adapt, grow, and fully develop their roles over time, ensuring long-term effectiveness beyond the initial team configurations. Only through this approach can a system truly stabilize and thrive, creating a new organic structure that functions as an integrated whole.

We'll dive deeper into this concept in later sections.

Gall's Law and Large Projects: Why Complexity Defies Predictability

Gall's law teaches us that *all complex systems that work evolved from simpler systems that worked*. This principle underscores a critical truth about large, multi-quarter projects that many CEOs don't want to hear: The complexity of these projects grows so large that the full scope of their timeline, tasks, and dependencies becomes unknowable.

The more complex a project, the harder it is to accurately predict outcomes, leading to missed deadlines, budget overruns, and escalating frustration. The fundamental problem lies in the very nature of complexity. A large multi-quarter project isn't just a collection of tasks—it's an intricate web of interdependencies, unknown variables, and emergent challenges.

If you include Brooks' law in this equation, as the size and scope of the project grows, the **interactions between components multiply exponentially**, making it nearly impossible to foresee all potential issues. Even with the best planning, unexpected obstacles—technical, organizational, or market-driven—will arise, further throwing the project off course.

Why Large Projects Fail: The Unknowable Complexity

When a project becomes too large, its complexity resembles that of a poorly designed system: one where the interactions between subcomponents are too numerous and intertwined to manage effectively. Tasks that seem predictable on paper often diverge due to unforeseen interactions.

Here are some common failures we see in large project development:

- **Dependencies cascade:** Completing task A depends on task B, but task B is delayed because it depends on task C, and so on. Work becomes frozen because the rigidity between the dependencies is too tightly coupled.

- **Emergent behaviors:** New problems arise from the interaction of project components that were not, and could not, be anticipated during planning. These emergent behaviors start to boil.

- **Compounding risks:** The longer the project timeline, the more likely it is that external factors (market changes, resource shortages, etc.) will disrupt progress.

The way to combat these challenges isn't to tackle the project as a monolithic whole but to break it into smaller, well-functioning subprojects and subsystems with well-defined correlations and clear boundaries.

Breaking Down Complexity: Subprojects and Subsystems

Extending Gall's law, we see that complex projects must be composed of simpler, well-functioning subprojects. Each subproject or subsystem must be designed to work independently while contributing to the larger whole.

This approach lowers complexity and makes timelines and budgets more predictable because:

- **Iterative progress ensures functionality:** When subsystems are built, tested, and refined incrementally, you reduce the risk of encountering major issues late in the timeline.

- **Simpler systems are easier to manage:** Smaller subprojects allow for more accurate scoping and better adaptability to changes or setbacks.

- **Boundaries reduce complexity:** By creating clear boundaries between subprojects, you minimize the interdependencies that cause cascading delays.

- **Better modularity:** By ensuring that each subsystem works independently, you create a foundation of smaller, reliable components that can be combined to achieve the larger goal.

The Case for Iterative Growth in Projects

"But I need to know when we're going to be done! Our board and our customers are pressuring us!" says almost every CEO everywhere.

Human nature drives us to want to simplify complexity and unconsciously attempt to distill intricate projects into linear, manageable tasks. While this inclination makes us feel more in control, it doesn't align with the realities of large, complex systems.

Just because we desperately want complex situations to be understood in simple terms, it doesn't mean that they can be.

Under pressure from leadership, companies often cave to the CEO's demand to "determine a timeline," leading to rigid, waterfall-style plans that map out every milestone in exhaustive detail. While this creates the illusion of control, it does nothing to reduce the actual complexity of the project. Unforeseen dependencies, evolving requirements, and emergent challenges will still arise, forcing teams to scramble to meet arbitrary deadlines. Instead of bringing clarity, these plans introduce rigidity, waste effort, and ultimately slow delivery—proving that complexity cannot be scheduled away.

These efforts aim to tame the unknowable, but, in reality, they fail to reduce the actual complexity overall. Instead, the project becomes immediately frozen when we need it to remain liquid for the best chance of success.

Ironically, these months would be more effectively spent iterating through smaller, manageable subprojects that incrementally reduce complexity and provide clearer insights into the overall goal. Simplifying our understanding doesn't simplify the problem; only deliberate action, aligned with iterative development, can do that.

A multi-quarter project that attempts to "plan everything perfectly" from the start is doomed to fail because it ignores the unknowable complexity that emerges in large systems. Instead, projects should adopt an iterative approach, focusing on delivering smaller, functional increments that reduce uncertainty at every stage.

For example, instead of tackling a full product overhaul as a single project, break it into sub-projects such as:

- Design the underlying key features or components first.
- Focus on the workflow of how these components come together second.
- Migrate specific parts of the system incrementally.
- Introduce pilot programs to test new processes or technologies before scaling them.

Each sub-project becomes a contained complex adaptive system (CAS) itself, with its own goals, boundaries, and manageable complexity.

This approach allows the system to evolve in smaller, controlled steps, ensuring that the four principles—Speed, Stretch, Shield, and Sales—emerge naturally:

- **Speed** increases because smaller, self-contained systems reduce dependencies and shorten feedback loops.

- **Stretch** develops as teams gain experience, adapting to the learnings from each subproject.

- **Shield** is reinforced because each increment allows teams to recognize, surface, and address risks early, before they cascade.

- **Sales** improves as teams demonstrate progress in tangible steps, building internal confidence and external trust.

Together, these subsystems work in concert, creating a flow that leads to more predictable, resilient outcomes across the larger project.

Building Predictable Systems Through Simplicity

Gall's law is a reminder that the foundation of any large endeavor must be simplicity. Leaders who attempt to force success by adding more complexity to an already massive project will only accelerate its breakdown.

Instead, the path to success lies in creating small, loosely coupled, highly aligned functional systems that evolve organically to deliver on the goals of the more complex mission.

By embracing this principle, organizations can create projects and systems that are not only functional but also resilient, adaptable, and scalable. The result? Large, multi-quarter initiatives that don't boil over or freeze but instead flow forward with clarity and purpose.

Chapter 23:
System Boundaries in a Complex Adaptive System

As we previously discussed in Part 2 regarding the effective management of boundaries, boundaries are the invisible structures that define the interactions, responsibilities, and flow of information within a complex adaptive system (CAS).

Much like the walls of a container guiding the movement of liquid, boundaries create the framework through which teams, processes, and departments interact. When thoughtfully designed, these boundaries provide clarity and enable a system to operate in a liquid state, where adaptability and alignment coexist.

In the context of organizational systems, boundaries serve as a critical means to allowing all four sentinels to flourish. They allow Speed to rapidly deliver technologies according to the demands of the business. They enable Stretch to flourish by granting teams the autonomy to innovate and adapt. They empower Shield to protect the organization by maintaining stability and preventing chaos. They activate Sales to boldly go where the company has not gone before in business development.

These roles make boundaries a cornerstone of effective system design, balancing the needs of the business with the sustainable ability to deliver on those needs.

In a liquid state, boundaries strike the ideal balance. They provide enough structure to ensure alignment and stability while allowing the flexibility needed for teams to experiment, grow, and adapt.

Leaders play a pivotal role in designing, communicating, and maintaining effective boundaries. They identify the interdependencies within the organization and define clear areas of ownership to minimize overlap and friction. At the same time, they should encourage open channels of communication and collaboration, ensuring boundaries don't become barriers. Regularly reviewing and adjusting these boundaries is essential, as organizational needs and priorities evolve over time.

Maintaining a well-regulated system requires vigilance and adaptability. When boundaries are thoughtfully designed, they act as the connective tissue that holds the system together, enabling it to remain flexible and aligned even in the face of complexity. But boundaries alone are not enough. The most well-structured teams can still falter if the invisible forces shaping behavior within those boundaries are overlooked.

Culture as the Invisible Influence

Within any organization, culture operates as an invisible yet powerful influence that shapes how people behave, interact, and make decisions. While boundaries define what teams *should* do, culture determines how they actually work within those boundaries—whether they collaborate effectively, prioritize the right things, and uphold quality standards when no one is watching. While organizational culture is often spoken about in terms of values or mission statements, much of it is unspoken. Culture is passed down not through official policies or handbooks but through the behaviors and habits that people observe and imitate from their peers and leaders. Culture is essentially made up of the rules you can't write down—the subtle cues and social norms that guide how work gets done and how people relate to each other.

At its core, culture is what bridges the gaps in formal rules. It informs how team members should act when situations aren't explicitly covered by guidelines. This is crucial in complex environments because not every rule can be written down or defined. Imagine trying to document every possible decision-making scenario in a fast-paced startup or a large enterprise. The effort would be futile because there are too many variables at play, from personal interpretations to market conditions. Culture steps in as the framework people rely on to navigate these gray areas, offering a sense of shared understanding about what's valued and expected.

People learn cultural rules by copying behaviors, which is how culture perpetuates itself. New employees might not initially know how to handle informal processes, such as the company's approach to handling client feedback or the unspoken expectations around deadlines.

But over time, they absorb these norms by watching how their peers act in meetings, how leaders respond to challenges, and even how small decisions are made day to day. Just like in any other social group, people learn by observing the successful behaviors of others.

This cultural copying is especially critical when signals are too complex to define. In an ideal world, we could document all cultural expectations and have people simply follow them.

However, in a CAS, the environment is always changing, multiple rules interact with each other to produce new situations, and the rules are often too nuanced to articulate clearly. This is why imitation and modeling play such a huge role in building and sustaining culture. People see how decisions are made, how conflict is handled, and how priorities are set, and they begin to model those behaviors.

In this sense, culture acts as an invisible system that keeps teams aligned without the need for constant oversight. It smooths out communication and decision-making across the organization, ensuring that, even in the absence of direct instruction, people know how to behave. It's a system of shared understanding that evolves over time and reinforces itself through imitation, creating a cohesive environment where teams can function fluidly within established but often unspoken boundaries.

Is your company culture influencing the complex system within your business to normalize toward the liquid state, or is it locking you into boiling or freezing conditions?

Chapter 24:
How Orders of Complexity Evolve

In any organization, complexity evolves as the business grows, shifting from simple systems of individuals to more intricate layers of teams and departments. This transition can be understood through orders of complexity, a concept that helps explain how units within a system become more interdependent and structured over time.

In a lower order of complexity, individuals are treated as basic, interchangeable units, much like atoms in the natural world. As the organization matures, these individuals are grouped into teams; this forms the higher order of complexity, where the focus shifts from managing individual people to managing the interrelations between teams.

Lower Order: People as Units

At the base level *lower order of complexity*, individuals within an organization are considered as the primary units. Even though we know that no two people are identical, when organizations create rules or policies they often treat all individuals as if they were the same.

An example of this would be HR guidelines, leave policies, or compliance regulations.

For instance, a company's vacation policy applies equally to everyone, regardless of individual job roles or personalities. This is a necessary simplification to manage a diverse workforce, allowing leadership to create structures that account for the broadest possible range of needs and scenarios.

In this *lower order of complexity*, the organization operates by focusing on what all people have in common, reducing variability by applying blanket rules and policies. This is essential when a company is small, as fewer moving parts means that managing complexity doesn't require layered structures. However, as the company grows, treating individuals as identical units becomes inadequate.

Imagine a company with a fixed policy of 20 vacation days per year for all employees, which must be taken as full days. While this may seem fair on the surface, as the company grows and teams expand, coordinating time off becomes increasingly complex. A parent with young children may need to take scattered days or hours off for school events or a sick child, while another employee might want to cluster their time off for international travel.

As more employees join with diverse needs and schedules, the rigid policy begins to strain both individuals and teams. Managers struggle to balance workloads when multiple key contributors are away, and employees often feel their personal needs are at odds with company norms.

What once worked for a small team now risks creating frustration and disengagement across a larger, more diverse workforce. The uniform approach sacrifices flexibility for simplicity, highlighting the growing challenge of balancing individual needs with operational efficiency.

Higher Order: Teams as the New Unit

As organizations expand, the system evolves from managing individual units to managing teams, which is a *higher order of complexity*. Teams are made up of individuals, but once they're grouped together, the dynamics change.

A new layer of complexity emerges because teams aren't just collections of people; they develop unique ways of functioning, forming their own goals, processes, and internal cultures. When teams become the dominant unit of organization, the focus shifts from managing individuals to managing team interactions and how those teams contribute to the organization's overall mission.

Just as in nature, where molecules are composed of atoms, teams are formed from individuals and begin to function as independent entities. This higher order of complexity requires a different approach: Rather than treating every individual the same, organizations now have to account for the interactions between teams and the different functions they perform.

For example, a development team and a marketing team might have very different workflows, goals, and challenges. The rules that govern them must accommodate these differences while ensuring alignment with the larger organizational structure.

Interaction Between Layers: Signals and Copying

There are two main ways to manage the interaction between different orders of complexity for both individuals and teams. These methods are *signals* and *copying*:

- **Signals:** In simpler systems, individuals or teams often align around a common goal or a set of processes, which serve as signals that guide behavior. For example, clear KPIs or project deadlines are signals that keep team members focused on shared outcomes. Everyone follows a standard path toward achieving a common objective, much like how atoms in a molecule respond to shared energy forces.

- **Copying:** When signals become too complex or nuanced, individuals and teams learn by copying the behaviors of those around them. This is particularly important in the context of organizational culture, where the unspoken norms of interaction and collaboration can't always be written down. As teams evolve, they absorb these cultural elements from more experienced members or through observing leaders, ensuring the transfer of values, processes, and behaviors even in complex environments.

By understanding the orders of complexity within an organization, leaders can effectively guide the transition from individual management to team management.

As the system evolves, the focus shifts from treating individuals as interchangeable units to recognizing the distinct roles teams play within the larger structure. To support this shift, leaders define boundaries that clarify which decisions teams own independently and where alignment across groups is necessary. These boundaries help reduce friction while ensuring collaboration happens with intent rather than by default.

Signals play a critical role in reinforcing these boundaries. Leaders amplify signals that highlight company-wide priorities and suppress those that create unnecessary noise between teams. Equally important is modeling boundary-respecting behavior, which demonstrates when to escalate decisions and when to trust teams to operate autonomously. When boundaries, signals, and behaviors align, teams gain the clarity and flexibility to function effectively while remaining connected to the organization's broader goals.

Liquid

Chapter 25:
Team Structure—
The Quark Analogy

Teams in a growing organization don't function in isolation. Their structure and interdependence are critical to maintaining balance and flow within the system. A team that lacks effective integration with other teams will struggle to contribute effectively to the organization.

To illustrate this, we turn to physics, where fundamental particles known as quarks provide a powerful analogy for team dynamics. Quarks are the building blocks of matter, but they never exist alone. They always come in tightly bound groups, held together by powerful forces. Separating a quark from its group takes so much energy that it doesn't simply break away—it causes new quarks to form.

Their interdependence is absolute. In the same way, teams within an organization aren't independent units but interconnected forces that rely on each other to function effectively. Just as quarks work in coordinated groups to form stable matter, teams must work together, with clearly defined roles and dependencies, to build a stable and scalable organization.

This analogy helps us understand how pulling teams apart arbitrarily can lead to dysfunction and inefficiency, while a well-structured system enables each specialized role to contribute to the overall effectiveness of the organization.

The Quark Analogy Explained

Let's dive deeper into the similarities between quark behavior and organizations. Considering that an organization, by definition, consists of more than one individual, we say that in organizations, individuals cannot operate in isolation. They must be part of a well-structured team with complementary skills and roles.

For example, a software development team is composed of developers, testers, DevOps engineers, product managers, and team leads, each playing a distinct role in the team's overall performance. Just as quarks interact and depend on one another to form stable particles, team members rely on each other to achieve their collective goals.

If you tried to pull one person out of a team and expected them to perform the same functions alone, their effectiveness would collapse. A single developer without the support of a DevOps engineer or product manager might quickly become overwhelmed by tasks that fall outside their core expertise.

Similarly, a team that lacks a leader or a product manager may find it difficult to make decisions, leading to inefficiency and disorganization. Teams need a **specific composition** to function, and this composition is driven by the activities the team is chartered to carry out.

Each role is essential, and without the right balance, the team ceases to be effective—much like how quarks disintegrate when isolated.

The Importance of Specialized Roles

Teams aren't simply collections of interchangeable people. They consist of individuals who possess specific, non-fungible skills that together form a cohesive unit. This interdependency is what makes teams so powerful. The specialized interactions within a team, such as the collaboration between a software developer and a DevOps engineer, or between a product manager and a designer, create a dynamic that allows the team to function optimally.

If every person on the team had the exact same skills and could perform the same tasks, the need for coordination would evaporate and the team would lose its structure and purpose. This would be akin to a team "boiling" and breaking apart, with no real cohesion to bind its members together.

The strength of a team comes from its diversity of roles and the mutual reliance of its members on each other to achieve their desired outcomes. Developers need testing to ensure quality, product managers to set direction, and operations teams to deploy code reliably. Without these distinct roles, the team would either freeze in indecision or boil in chaos, unable to perform its duties effectively.

Why Monoteams Fail: The Steam Analogy

This is why monoteams—teams made up of people who are perfectly fungible and interchangeable (generalists)—tend to fail, as we saw when Theo created a monoteam in Chapter 21.

A team where every member can do exactly what the others can might sound efficient in theory, but in reality, it lacks the structure and interdependency that make teams highly productive.

When no one relies on anyone else for specialized skills, the team becomes like a cloud of steam. They're individuals working independently, moving in multiple, uncoordinated directions, with no cohesive force keeping them aligned. The team becomes disorganized and the system eventually breaks down as tasks become disconnected and priorities unclear.

In the classic Mythical Man-Month theory we explored in Chapter 21, adding more people to a project doesn't necessarily make it go faster, because each new person requires training, context, and alignment with the rest of the team.

If teams are poorly structured, adding more individuals simply compounds the complexity rather than solving it. Our quark analogy helps explain why: *People who are not part of a highly aligned, interdependent team will contribute to boiling rather than stabilizing the system.*

Building Teams With Effective Boundaries

Creating highly functional teams requires structuring them with the right ontological pattern—ensuring individuals are interdependent and need one another to function effectively.

Team formation isn't about assembling random individuals but about strategic planning to achieve the right mix of skills, roles, and interdependencies.

When teams are structured correctly, they become cohesive units with clearly defined boundaries, a shared purpose, and an aligned direction. This principle applies across all teams within an organization, not just engineering.

The analogy of quarks in particle physics is again useful here. Just as quarks must exist in specific combinations to form stable particles, teams require the right blend of individuals and expertise to operate effectively. A well-structured team fosters a liquid flow state, where tasks move seamlessly and goals are achieved with efficiency and synergy. In contrast, poorly structured teams can either boil in chaos or freeze due to misalignment and indecision.

As organizations evolve, complexity increases. Leaders must establish clear boundaries to ensure teams function effectively. These boundaries define the structure that balances autonomy with interdependence. Without them, teams can become entangled in complexity (boiling) or isolated and ineffective (frozen). Proper boundaries enable an organization to operate in a liquid state, encouraging flexibility, adaptability, and sustained performance.

Growth within an organization often requires restructuring teams, much like in physics when quarks are pulled apart, creating new quarks in the process. When organizations scale, simply creating a new team of all new people often leads to inefficiency and communication overhead. Instead, successful growth requires creating new teams through a blend of existing members and new hires, ensuring cultural continuity while integrating fresh perspectives. This mirrors how Theo, in our story, learned that forming entirely new teams from scratch failed, while integrating experienced members with newcomers enabled smoother transitions and greater operational efficiency.

Effective boundaries aren't arbitrary but emerge organically based on the needs of the organization and the relationships between its components. Thoughtful structuring reduces complexity while ensuring teams remain interdependent and cohesive.

When team boundaries are well-defined, organizations benefit in multiple ways:

- Teams absorb the company's culture and practices while maintaining focus on their specialized tasks.
- Organizational complexity is minimized, encouraging innovation and productivity.
- Teams align with broader company objectives while operating efficiently within their defined roles.

When they recognize the patterns that shape team effectiveness and set appropriate boundaries, leaders create an environment where teams function optimally. This ensures the organization remains resilient and adaptable, maintaining a liquid state where collaboration and performance flow naturally, free from unnecessary complexity.

By acknowledging these complexities and carefully setting boundaries, leaders ensure that their teams are both **independent enough to innovate** and **connected enough to collaborate** effectively. The teams within the company become *loosely connected and highly aligned.*

Through thoughtful boundary creation, organizations can achieve the right balance, leading to a state where teams can thrive by remaining in an adaptable, liquid environment that we often call *flow*.

Liquid

Chapter 26:
The Power of Continual Growth

It's important to tie the concepts of system boundaries, cultural alignment, and team dynamics to the overarching theme of continual growth, which is what the Stretch sentinel truly represents.

The Stretch sentinel focuses on the ability of an organization and each person within it to scale sustainably, ensuring that as complexity grows, the system remains fluid, adaptable, and resilient. It's not just about growing fast but growing in a way that allows the company to stretch without breaking.

A well-functioning complex adaptive system (CAS) naturally produces the CTO Levels sentinels (Speed, Stretch, Shield, and Sales) without forcing the system to its limits. When the system continually operates in a liquid state, with the right boundaries and ontological patterns in place, it can flex and expand with the needs of the business. Stretch reflects the system's ability to handle increased demands, new complexities, and additional layers of structure without collapsing under the weight of its own growth.

For the CEO, these sentinels are visible markers of success in an technology organization. A CEO can't see the inner workings of the complex system at play within each of the teams within their organization, the subtle interactions between teams, or the boundaries that keep the system in balance. However, what they can perceive is the outcomes: how quickly the organization delivers (Speed), how well it adapts and scales (Stretch), how securely it's protected from threats (Shield), and how effectively it drives internal and external alignment (Sales).

An effective executive leader ensures that these sentinels are healthy and growing across the company and within all teams, aligning the company's internal structure with the CEO's perception of forward momentum and success.

The Stretch sentinel is about managing complexity as the company expands, ensuring that growth isn't chaotic or overwhelming but consistent and structured. Continual growth, guided by strategic boundaries and a clear understanding of the secret world at work within its complex system, is what enables an organization to thrive in the long term.

Your role as leader is to nurture this growth, ensuring that the system remains flexible, adaptive, and aligned with the business's overall objectives. This is how you continually stretch the organization toward sustainable success.

PART 4:

Influence the System Toward Better Outcomes

Spoon Boy: Do not try and bend the spoon. That's impossible. Instead, only try

to realize the truth.

Neo: What truth?

Spoon Boy: There is no spoon.

Neo: There is no spoon?

Spoon Boy: Then you'll see that it is not the spoon that bends; it is only yourself.

–From *The Matrix*

Liquid

Chapter 27:
Alice Experiences Growing Pains

Alice's company is thriving more than either she or Theo had expected. Not only did they survive the pains of organizational growth (see Part 3), but their B2B product is gaining traction with major enterprise clients and the B2C platform continues to increase its loyal user base. But the relentless pace of growth over the last year has taken its toll, and Theo is feeling the long-term strain of trying to keep up. He settles into the chair across from Alice to discuss what's next, his face a mix of weariness and determination.

"Something on your mind?" Alice asks, pouring herself a cup of coffee. She could tell this wasn't a casual conversation.

Theo leans forward, clasping his hands together. "Yeah, there is. I've been thinking a lot about where we are as a company and... where I fit into all of it."

Alice raises an eyebrow, leaning back in her chair. "That sounds serious."

"It is," Theo admits. "When we started, it was just the two of us and a couple of developers. Back then, I could keep everything in my head—features, bugs, architecture, all of it. But now, with two products, close to 40 people in the engineering team, and a constant flood of new priorities, I feel like I'm stretched too thin. I'm constantly juggling management and technical strategy, and I'm not sure I'm doing justice to either."

Alice frowns. "Are you saying you want to step down as CTO and leave the company?"

"Not exactly," Theo says carefully. "I love this company, and I want to keep contributing. But I've realized what I'm truly passionate about, what I'm best at, is designing the architecture, planning for scalability, and ensuring the tech can handle whatever comes next. I'm not the right person for managing teams and scaling the organization. I think we need a CTO who can focus on building the leadership structure we need to grow, someone who has the skills to manage the operational complexity of cross-functional collaboration."

Alice nods slowly, absorbing his words. "That's... a big shift. I appreciate your transparency, and it makes sense. You've built an incredible foundation, Theo, and if focusing on architecture helps you, and the company, thrive, then I think it's the right move."

Theo exhales, relief washing over him. "Thanks, Alice. I've been wrestling with this decision for a while, but I believe it's what's best for the company. I'll stay as Architect and guide the technical vision while we bring someone in to take on the CTO role."

"Okay," Alice says, smiling. "Let's make it happen. We'll find someone who can take us to the next level, and in the meantime, we'll need to ensure a smooth transition."

Theo returns her smile, feeling a weight lift. "I really believe this will set us up for success. And don't worry; I'm not going anywhere."

As the conversation winds down, both Alice and Theo feel a renewed sense of focus and purpose. Theo's decision to step into the Architect role isn't a step back but a step toward aligning his strengths with the company's evolving needs.

It marks the beginning of a new chapter. This transition requires careful planning and the right leadership to guide the company into its next phase of growth.

Chapter 28:
Introducing the Sales Sentinel

As Theo transitions into the Architect role and Alice brings in a new CTO, the need for cross-functional alignment becomes increasingly evident. Their struggles with miscommunication, conflicting priorities, and uncoordinated departmental efforts reveal a critical gap: the need for effective internal and external "sales."

The **Sales** sentinel serves as the framework for bridging these divides. It emphasizes not just the traditional function of selling products to customers but also the internal selling of ideas, priorities, and vision within the organization.

For CTOs, this means actively influencing and aligning teams to ensure that everyone, from engineering to marketing to sales, works cohesively toward shared goals. By mastering the Sales sentinel, leaders can navigate the complex dynamics of organizational alignment and drive the company forward with unified purpose.

Liquid

Chapter 29:
Leveraging Signals

In a complex adaptive system (CAS) like a growing organization, leaders shape behavior, alignment, and performance by amplifying or suppressing signals.

Signals, the cues and messages that guide decision-making, are the primary mechanism through which leaders influence the system and steer it toward the desired outcomes. These signals dictate how information flows across the organization, helping teams prioritize tasks and adapt to changes.

As companies—and especially technology organizations—scale, the Sales sentinel becomes particularly reliant on effective signal management.

This sentinel measures how well the organization aligns internally while ensuring its external positioning remains strong with customers. The interactions between engineering, sales, and customer success teams are often guided by the clarity and consistency of these signals.

Effective leadership involves making intentional choices about signal amplification and suppression. Leaders must amplify clear, unified messages that align teams on shared objectives, ensuring everyone moves toward common goals.

Simultaneously, they must suppress noise—conflicting or irrelevant information—that might create confusion and disrupt focus.

Managing signals with care allows leaders to reduce misalignment, encourage collaboration, and create an environment where the organization can adapt and grow sustainably.

Chapter 30:

Three Types of Signal

Every organization operates as a system that's influenced by the constant flow of signals. These signals function as the language of the system, directing how individuals and teams respond to both day-to-day tasks and larger strategic objectives. Leaders who manage these signals effectively can shape their organization's flow to drive alignment and sustain momentum.

There are three primary types of signal that determine how work progresses and how teams align within an organization:

- explicit signals
- implicit signals
- feedback loops

Explicit Signals: Aligning Teams Through Directives

Explicit signals are the clearest and most direct form of guidance. These signals include instructions, goals, metrics, and performance targets. They establish expectations and serve as visible markers of success, providing teams with clarity on what's required.

When explicit signals are strong and consistent, teams know what's expected and can adjust their behavior accordingly.

Explicit signals within a company often come in the form of business goals, release deadlines, sprint plans, and uptime service level agreements (SLAs). For example, an engineering team may receive a directive to increase system uptime to 99.99% or lower their post-launch defect rate to under 2%. These explicit signals set a clear target that shapes daily operational decisions, encouraging investment in infrastructure reliability and incident response processes.

Implicit Signals: Leadership by Example

Implicit signals are more subtle and often unspoken. They're deeply intertwined with an organization's culture, shaping how teams understand what's valued and expected without explicit instruction. They shape team behaviors through observation, social cues, and cultural norms.

Mimicry and copying play a significant role in transmitting these signals. People naturally observe those who are in positions of influence and mirror their behaviors, adopting their approaches to problem-solving, collaboration, and decision-making. Leaders play a crucial role in transmitting implicit signals by modeling the behaviors and values they want to see reflected across the organization. When a leader demonstrates urgency, accountability, or collaboration, others pick up on these cues and align their actions accordingly.

Within a company, implicit signals might be conveyed when an engineering leader stays late to assist with a critical deployment. This action implicitly communicates that meeting delivery commitments is valued and that teamwork during high-stakes moments is expected. Similarly, when a CTO regularly engages with product teams during stand-ups, it sends a signal that cross-functional alignment is a priority.

Feedback Loops: Continuously Aligned Decision-Making

Feedback loops ensure that the system adapts and corrects itself over time. These signals emerge from the outcomes of work and inform teams about whether their actions are leading to the desired results. Effective feedback loops reinforce positive behaviors and prompt adjustments when necessary, helping the organization remain responsive to challenges and opportunities.

Feedback loops often manifest in postmortems after incidents, sprint retrospectives, or product performance reviews. For instance, after a failed production deployment, a postmortem can identify root causes and process gaps. This feedback loop informs future deployment processes, reducing the likelihood of similar failures and fostering a culture of continuous improvement.

Cross-functional feedback loops are also critical. For example, sales teams might relay customer objections or feature requests to product and engineering teams, enabling adjustments to the product road map. Similarly, customer support or services teams can provide feedback on recurring user issues, leading to improvements in usability, onboarding processes, or documentation.

These cross-functional feedback loops help align the entire organization around customer needs and ensure tighter collaboration between technical and business teams.

Chapter 31:
Signals as the Mechanism of Influence

Imagine the complex system within a business as a neural network, where nodes (teams and leadership layers) amplify or suppress signals to regulate the flow of information. Just as neurons manage impulses, leaders must carefully curate how messages circulate through the organization. The ability to regulate these signals—deciding which ones to suppress, which ones to amplify, and how to direct the latter—is a leader's primary mechanism for influencing the system. Effective leaders ensure the right information reaches the right people while unnecessary noise is filtered out, preserving clarity and focus.

Consider the example of a CTO managing a strategic workforce reduction. The full scope of the decision may involve complex analyses of individual contributions, future project needs, and organizational priorities. However, when this decision is communicated to direct reports, the CTO simplifies the message to focus on the rationale for the decision and its immediate implications.

By the time the message reaches individual employees, it has been distilled further: "We no longer need your services." This multilayered filtering reduces the complexity for each audience, ensuring clarity while maintaining coherence across the organization. This filtering isn't just a practical necessity; it's essential for organizational health.

A leader attempting to communicate every detail to every individual risks overwhelming the system with noise, leading to confusion, inefficiency, and misalignment. Regulating signals ensures that each layer of the organization understands what's most relevant to them, focuses attention on key priorities, and reduces unnecessary complexity. By carefully managing these signals, leaders cultivate coherence and alignment within the system, empowering teams to act with purpose and precision.

The Role of the Right-Hand Person

When teams are small, a CTO often has the capacity to directly oversee all of the work, maintaining close oversight and direct communication with each individual. However, as the organization grows, this direct oversight approach becomes unsustainable. The increased size and complexity of the team introduce more layers of communication and coordination, making it nearly impossible for the CTO to directly regulate every signal.

To regulate signals effectively in a complex and growing organization, a CTO must rely on a right-hand person, which is typically a VP of Engineering, a Development Director, or a similar role. This individual becomes an essential conduit for the CTO's signals, translating strategic vision into actionable guidance for the broader team. They ensure that messages are clarified, appropriately amplified or suppressed, and strategically propagated throughout the organization.

Without this intermediary, as the company grows the CTO risks becoming overwhelmed, trapped in the weeds of daily communication and decision-making. Delegating this responsibility creates a scalable system where the right-hand person acts as a crucial node, both distributing key information and maintaining continuity. They ensure that the organization remains coherent even in the CTO's absence, while allowing the CTO to focus on strategic priorities.

The value of this role cannot be overstated. A capable right-hand person understands the company's culture, goals, and operational rhythms. They become a critical boundary within the organization, shaping how signals flow through teams and ensuring alignment with the organization's needs. This individual doesn't merely transmit messages; they refine them to suit their audience, balancing clarity with relevance and maintaining the system's overall health.

Theo's transition into the Architect role illustrates how a well-placed right-hand person can strengthen signal regulation within a growing organization. With his narrowed focus on technical architecture and long-term scalability, Theo will be able to amplify signals that prioritize system resilience, such as investments in infrastructure upgrades, modular code design, and proactive capacity planning. He suppresses signals that pressure the team into rapid, unsustainable feature releases, ensuring that technical debt doesn't spiral out of control. Meanwhile, the new CTO will be able to focus on day-to-day operational excellence and cross-departmental alignment. They can amplify signals that emphasize collaborative planning, like hosting biweekly project review meetings with product, engineering, and sales leads to ensure road maps are synchronized. They will also conduct monthly bottleneck reviews, identifying friction points in product delivery and coordinating solutions across teams. By splitting these responsibilities, Theo and the incoming CTO will ensure that both technical stability and cross-functional execution remain in balance as the organization scales.

Theo's decision to step into the Architect role isn't a retreat but a deliberate redefinition of system boundaries. He sends explicit signals by establishing clear architectural principles and setting targets for system performance, such as requiring all new services to meet scalability benchmarks before deployment.

He also sends implicit signals designed to assist in establishing the new CTO: deferring operational decisions to them in leadership meetings, publicly supporting their initiatives, and redirecting technical questions from other executives to the CTO. Through these implicit signals, Theo reinforces the new CTO's authority and his own role as the new operational leader. This example underscores the importance of specialized leadership roles. By delineating responsibilities and creating clear boundaries, organizations can allow signals to disseminate effectively, ensuring a balanced, liquid state where teams remain aligned, adaptable, and capable of thriving amid complexity.

Coherence and Decoherence: Balancing Signals Across Teams

In a growing organization like Alice and Theo's, signals (messages, directives, and information) act as the connective threads that translate the company's goals and vision into actionable daily operations. The challenge lies in managing these signals to achieve a delicate balance: encouraging **coherence**, where teams align on shared objectives, while embracing **decoherence**, which allows teams the autonomy to innovate and specialize. Striking this balance is essential for maintaining organizational flow, where each team contributes meaningfully to collective goals without becoming entangled in unnecessary complexity:

- **Coherence** ensures that everyone in the organization, regardless of their specific focus, understands and works toward the company's vision. It aligns teams under shared priorities and objectives.

- **Decoherence**, on the other hand, empowers teams to achieve those objectives in ways best suited to their domains, promoting adaptability, innovation, and efficiency.

Together, coherence and decoherence create a dynamic equilibrium where the organization operates as a cohesive whole yet remains flexible enough to thrive in the face of complexity. For Alice and Theo, the decision to separate their teams into B2C and B2B units exemplifies this balancing act. The B2C team was tasked with enhancing user experience while the B2B team focused on enterprise-specific requirements. This division fostered coherence by aligning each team's work with the broader company vision, and introduced intentional decoherence by granting each team the independence to tailor solutions to their unique objectives. This separation reduced overlap, minimized cross-team distractions, and clarified responsibilities, enabling both teams to focus on their distinct goals.

Signals are the mechanism that makes this balance possible. The CTO plays a pivotal role by amplifying the right signals, such as the company's overarching priorities and long-term vision, while suppressing those that could create noise or distractions. The CTO ensures that the B2B team isn't burdened with consumer-focused updates and shields the B2C team from enterprise compliance details. By directing the flow of signals, Alice and her leadership team enable each group to remain focused on their goals while contributing to the company's unified strategy. By carefully managing this interplay of signals, Alice and Theo have set their organization on a path to sustainable growth, ensuring that teams remain aligned with the company's strategy while retaining the freedom to innovate and execute with precision.

Simplifying the Organization Through Signals

For Alice and Theo, simplifying the organization didn't mean reducing it to its simplest parts. Instead, it required properly managing the flow of signals to maintain clarity, purpose, and alignment while minimizing unnecessary complexity. The division into B2C and B2B teams marked an important first step, but the true challenge lay in ensuring that signals were being broadcast effectively and meaningfully across all layers of the organization.

Simplification begins with leaders establishing clear boundaries where signals can be amplified or suppressed as needed. For example,

Alice worked with her leadership team to amplify high-priority messages, such as the company's goals for market expansion, ensuring these signals reached both teams. At the same time, details about specific customer implementations or team-specific challenges were confined to the relevant groups, reducing noise and keeping the broader system focused.

At the core of this approach lies the company's vision and goals, which act as the guiding signal for all activities within the system. Leaders like Alice, her executive team, and the new CTO must ensure that this signal remains clear, consistent, and compelling as it moves through the organization. At the same time, they must suppress signals that could add noise or misalign priorities, such as redundant processes or conflicting departmental objectives.

When signals are managed in this way, the organization naturally simplifies itself—not by cutting corners but by ensuring each team receives the information necessary to contribute effectively. This approach provides the foundations for a liquid flow state where teams are aligned with the big picture yet are free to innovate and execute in ways that suit their unique responsibilities.

For Alice and Theo's company, this balance proves essential to maintaining momentum and achieving sustainable, scalable growth. Alice is soon to see that managing signals isn't something she can do sporadically; it has to be a consistent habit.

Liquid

Chapter 32:
Alice's Leadership Team Hits a Breaking Point

Alice leans back in her chair, exhausted but determined. After weeks of interviews and careful deliberation, she has hired Jordan as the company's new CTO. Jordan has an impressive track record, with years of experience scaling engineering teams at rapidly growing startups, a reputation for driving innovation, and a sharp understanding of operational processes. For Alice, this hire is a significant milestone. With Theo transitioning to the Architect role, she needs someone who can take the reins of the engineering organization and ensure the company's technical and operational scalability.

Jordan's arrival brings fresh energy and optimism. He quickly identifies areas for improvement, implementing streamlined processes and introducing new frameworks for cross-functional collaboration. He speaks confidently about building a culture of innovation and efficiency. The team is invigorated by his presence, and Alice feels a wave of relief. Maybe, just maybe, things are finally falling into place. The company's progress surges, but the honeymoon period doesn't last.

The cracks begin to show during a leadership meeting just a few months later. Alice calls the meeting to align the executive team on the company's aggressive sales goals for the next quarter.

The marketing lead argues for increased resources for international campaigns, while the head of customer success pushes back, citing concerns about strained support capacity.

Jordan, meanwhile, expresses frustration that engineering deadlines are being derailed by last-minute changes from sales. The meeting devolves into a heated debate, with each department head defending their own priorities.

Alice tries to mediate, but it's clear that the team is working at cross-purposes. The once-cohesive leadership team has become a collection of silos, each operating in isolation and prioritizing its own agenda over the company's shared vision.

After the meeting, Alice sits down with Jordan to debrief. "That was a mess," she says, shaking her head. "Everyone's pulling in different directions. I don't even know how to bring them back together. We were doing so well; how did this happen?"

Jordan sighs. "It's not uncommon at this stage of growth. The bigger we get, the harder it is to keep everyone aligned. But we can't keep going like this. If we don't fix the communication and accountability issues at the leadership level, it's going to trickle down and impact the entire company. Trust is essential to our continued success."

Despite Jordan's warnings, the situation worsens over the next few weeks. Miscommunication between departments leads to stalled initiatives and mounting frustrations. Sales sets aggressive goals that engineering can't support within the existing timelines. Marketing launches campaigns without consulting customer success, overwhelming the support team. Even Jordan, with all his expertise, finds it difficult to manage the growing tensions.

Alice feels the weight of it all. She has always prided herself on being a decisive leader, but now she's doubting every move she makes.

Advisors offer conflicting advice. Some suggest a more hands-on approach, while others push for delegation or firing the worst-offending leaders and replacing them with people who would "get in line."

Meanwhile, the executive team's lack of trust and alignment becomes more apparent with each passing day. After weeks of mounting tension, Alice calls Jordan into her office. The executive team's misalignment isn't just a nuisance anymore; it's a threat to the company's future. She needs a way to get everyone on the same page and working toward the same goals. Despite her mounting stress, which usually leads her into command and control behavior, she instead decides a more inclusive approach might be effective.

"Jordan," she begins, "this can't go on. We're losing time, and every department seems to have a different idea of what's important. I need your help to fix this. You have extensive experience working at scale. What have you seen work at other businesses when they hit this inflection point?"

Jordan nods. "I've seen this before. At this stage of growth, it's common for departments to operate like islands. We need a framework to tie them together. We need a system that aligns priorities, clarifies accountability, and ensures everyone understands how their work impacts other parts of the company and how it contributes to the company's overall vision. We need a system that forces us into better alignment through more consistent feedback loops."

Alice leans forward. "This new system sounds interesting. Okay, where do we start?"

Jordan proposes implementing a business operating system (BOS) to create structure and transparency across the organization. They decided on a hybrid approach, borrowing elements from systems like objectives and key results (OKRs),[18] goals, experiments, and measures (GEMs),[19] and the Entrepreneurial Operating System® (EOS).[20]

The goal is to establish a shared framework that will drive alignment while remaining flexible enough to adapt to the company's unique needs.

Their first step is to align on the vision and goals for the next 12 months. Jordan suggests a two-day leadership offsite to define the company's overarching objectives and identify the key results they need to achieve. "This isn't just about creating goals," Jordan explains. "It's about creating a shared language for success—something every department can rally around."

Alice loves the idea and organizes the leadership retreat. Within a month, everyone's gathered at a ski lodge in Winter Park, Colorado, for two days of realignment.

The offsite, however, isn't easy. Tempers flare regularly as department heads disagree on priorities. As the facilitator, Jordan handles the flare-ups with world-class finesse and, rolling into day two, they manage to identify three core objectives that align with the company's long-term strategy. From there, they break down how each department contributes to these objectives, defining key results that are specific, measurable, and achievable.

With a plan in place, Alice and Jordan turn their focus to the operational structure. They outline a new executive meeting cadence to ensure consistent communication across the leadership team. Weekly check-ins are implemented to review progress on their core objectives, while quarterly planning sessions allow for course corrections as needed. To address accountability, they introduce RACI charts (responsible, accountable, consulted, and informed) to clarify roles and responsibilities for cross-departmental initiatives.

"This is how we make the system work," Jordan explains. "Everyone knows what they're responsible for, and we have a regular rhythm to ensure we stay aligned." The BOS cadence ensures each part of the company stays in alignment with the others and ensures everyone has to work together in order to achieve their goals.

Finally, they address communication flow. Jordan leans forward, his tone firm but measured. "When everything is a priority, nothing is a priority," he says. Since stepping into the CTO role, he's seen how easily competing demands can throw the team into chaos. He points to a recent incident where conflicting priorities from the sales and engineering teams caused confusion around a critical feature release. Sales had promised a delivery date to a key client, while engineering, unaware of this commitment, had deprioritized the feature in favor of resolving platform stability issues.

"If we amplify every signal," he says, "we'll overwhelm the system. That's exactly what happened last month when everyone was trying to push their priorities at once. Sales was escalating directly to developers, product was pushing for changes, and engineering was already scrambling to meet the security audit deadline. People were getting pinged constantly, and no one knew what to prioritize."

He continues, "If we suppress too much, we'll lose critical information. Remember when the customer success team raised concerns about that enterprise client's onboarding issues, but it never made it to product until we were already at risk of losing the renewal? That's what happens when the right signals get buried."

Jordan concludes, "If we amplify every signal, we'll overwhelm the system. If we suppress too much, we'll lose critical information. We need to create boundaries that ensure the right signals reach the right people at the time that they need that information. I propose we create a weekly priority alignment meeting with product, sales, and engineering leads. That way, we can surface urgent cross-team needs early and make sure everyone knows what matters most before people start their week."

Together, Jordan and Alice work to define communication boundaries at every level of the company. The executive team focuses on high-level strategy, department heads cascade relevant signals to their teams, and individual contributors escalate issues through well-defined channels. This structure not only streamlines communication but also reinforces trust across the organization. The new system is rolled out incrementally throughout the organization, and the change is palpable. Leadership meetings become more focused, with clear agendas and actionable outcomes. Teams are energized by the clarity of their goals and the knowledge that their work is directly contributing to the company's success. For the first time in months, Alice feels like the company is back on track.

"Jordan," she says during a weekly check-in, "this BOS... It's a game-changer. I don't know why we weren't focused like this sooner."

This feedback loop between signal emission and team response proves essential for maintaining focus and sustaining flow. The same principle applies to sales, where the directive to achieve a specific revenue target acts as a signal, aligning the sales team's actions and priorities with the company's overall goals. When signals are clear, actionable, and consistently reinforced, they become reliable benchmarks against which teams can calibrate their efforts, ensuring alignment across the organization.

Culture as the Invisible Influence

While directives and role modeling are powerful tools for influencing teams, *culture* serves as the dominant signaling influence across the executive team and the organization as a whole.

Culture acts as the connective tissue bridging the boundaries between teams. While engineering prioritizes quality and timeliness and sales focuses on closing deals, the company's culture needs to align both groups around a shared vision of delivering value to customers. Culture, in this sense, functions as an implicit boundary shaping team behaviors, fostering alignment, and preventing unnecessary signals or distractions from crossing into unrelated areas.

For example, Alice observes that while the engineering team had cultivated a strong delivery-focused culture, this ethos had not yet fully taken root in the sales team. Recognizing this gap, she takes action to embody the culture she wants to see. Culture isn't something that can be dictated from the top; it must be demonstrated and lived by leaders, allowing others to observe and adopt it through alignment with shared values and behaviors.

Alice recognizes an opportunity to shift this mindset when she observes the Chief Revenue Officer (CRO) struggling to coach their team. While the team achieves some targets, they lack a structured, repeatable process for consistently closing deals and delivering results. As CEO, Alice understands that issuing directives alone won't solve the issue. She needs to model the behaviors she wants to see embedded in the team, so she partners with the CRO to build a culture of disciplined execution.

Alice begins by joining key sales leaders on strategic calls with high-value clients. During these calls, she demonstrates her approach: deeply understanding client needs, articulating the company's value in terms of long-term benefits, and emphasizing clear, actionable next steps. Following each call, Alice debriefs with the CRO and sales leaders, highlighting what worked and brainstorming ways to refine the process for the broader team.

In private discussions with the CRO, Alice reinforces the importance of structure and consistency. "Every call is a step in a larger journey," she explains. "We need a clear, repeatable process for identifying client needs, demonstrating value, and following through. A strong process leads to reliable results."

By actively engaging in sales activities and partnering with leadership, Alice demonstrates how to align effort with outcomes. Her example sets the tone for a disciplined, value-driven culture within the sales organization. Over time, the CRO incorporates these practices into the team's workflow, creating a more structured and delivery-focused sales process that aligns seamlessly with the company's broader objectives. Through her partnership and visible leadership, Alice ensures that the sales team becomes a reflection of the company's commitment to consistent, meaningful results

Alice works intentionally to model the desired cultural traits—discipline, focus, and customer-centricity—through her interactions with the sales team and her broader leadership approach. By consistently reflecting these principles, she allows the culture to take shape organically, leading to alignment across the organization. Over time, the teams absorb and internalize these behaviors, creating a unified culture that guides decision-making and strengthens the company's shared vision. Culture, when intentionally lived and reinforced by leadership, becomes an invisible yet powerful signal, shaping the organization from within. It defines how teams align, interact, and deliver on their objectives, ensuring that the company's vision flows seamlessly through every level.

Utilizing Signals to Align Leadership

As the company grows, Alice and Jordan begin to see a new challenge emerging. The executive team, once a tightly knit group navigating the early stages together, is now showing signs of fragmentation. Functional leaders are optimizing for their own departments, but cross-functional alignment is beginning to fray. Sales is driving aggressive commitments to customers, while engineering is feeling the strain of those promises. Marketing is pushing ambitious campaigns, but customer success is warning that the support team may not be able to keep up with the influx of new clients. The seams are starting to show.

Alice and Jordan recognize that influencing their peers on the executive team requires a different approach than guiding their individual teams. While signals and behaviors work effectively within functional groups, aligning the executive team demands something more: a shared system to connect the leadership team's priorities around the company's broader goals. Without this alignment, they risk teams operating in silos, undermining the collective progress of the business.

To address this, Alice and Jordan introduce a BOS to bring consistency and clarity to cross-functional decision-making. The BOS becomes the foundation for their executive alignment. It creates a structure where every leader has visibility of key objectives, dependencies, and progress. It promotes transparency, ensuring that when sales pushes for faster delivery, engineering can openly raise capacity concerns and marketing can adjust expectations accordingly.

Within this structure, the Sales sentinel becomes more than an external tool for driving revenue. It becomes a cross-functional alignment mechanism. Alice works closely with the CRO to ensure sales commitments align with engineering's capacity and product's development timelines. Together, they establish a shared understanding that the success of the sales team cannot come at the cost of delivery failure. Every team's success is interconnected.

Alice begins to see that her primary lever in this new reality isn't exerting more control over each team but instead managing the signals that shape behavior across the system. Explicit directives, such as deadlines and key metrics, are critical, but so are the implicit cultural norms that guide how teams work together. Alice amplifies signals that reinforce the company's shared focus on delivering customer value while cutting down on noise—like competing priorities and last-minute shifts—that could pull teams out of alignment.

A practical example of this emerges when Alice and Jordan establish a feedback loop between sales, product, and engineering. Sales gathers customer insights from the field, which are then filtered and prioritized through product. Engineering receives this refined input early, giving them time to adjust their plans proactively. This loop prevents the disruptive pattern of sales making late-stage promises that derail product road maps or force engineering into firefighting mode. Managing these signals ensures that the right information flows to the right teams at the right time. Alice and Jordan begin to see their work less as controlling outcomes and more as shaping conditions. They're orchestrating flow—using the BOS, the Sales sentinel, and signal management to guide the organization toward a state where teams are both aligned (coherence) and free to execute independently (decoherence). This liquid state allows the company to adapt and thrive as complexity increases.

In this phase of growth, Alice's role isn't to fix every problem herself but to ensure the system can solve its own problems. She shapes the environment, ensuring the teams have the clarity and autonomy they need to adapt. This is leadership within a CAS—less about directing and more about cultivating the conditions for sustained success.

Liquid

Chapter 33:

The Power of Redundancy

As Alice and Jordan shape the environment for alignment and flow across the executive team, they come to understand that sustaining this balance requires more than signal management and goal-setting. A resilient organization cannot rely solely on alignment; it must also account for the inevitable uncertainties of growth. Teams will face turnover, unexpected challenges, and shifting demands.

To remain adaptable under these pressures, the system must possess both redundancy and specialization—two forces that, when balanced, ensure the organization is prepared to withstand disruption while continuing to innovate.

In a complex adaptive system (CAS) like a growing organization, redundancy and specialization are essential elements for maintaining resilience and adaptability. These concepts allow the system to thrive by ensuring no single failure cripples the organization while simultaneously fostering unique capabilities that drive innovation.

Redundancy and specialization work together to create a balanced foundation for the Sales sentinel, which governs a company's ability to communicate and execute ideas effectively. Redundancy ensures there's enough overlap between roles to maintain continuity and resilience, while specialization allows each team or leader to focus on their unique strengths.

This dynamic interplay not only protects the organization from disruptions but also ensures critical strategies and messages are communicated clearly across teams and departments, enabling both internal alignment and external execution. Without this balance, a company risks either fragmentation or stagnation, undermining its ability to sell ideas internally and deliver value externally.

Avoiding Single Points of Failure

A key principle of redundancy in a complex system is to avoid "key person dependency," where the absence of a single individual can cause the entire system to falter. Consider Alice and Theo's company: If Theo were the only one capable of managing the technical architecture, his absence due to burnout or unexpected circumstances could collapse the system. Redundancy addresses this risk by ensuring critical functions aren't dependent on a single person.

This is why Alice and Jordan invest in additional leadership roles like a development director, a "right-hand person" who can propagate signals to individuals across the teams and can also step in for Theo or Jordan when needed.

While some might see this redundancy as inefficient, it's a necessary investment in the system's resilience as the company grows. By sharing the workload and creating overlap in responsibilities, the organization ensures more diversity of thought; this allows teams to find answers quickly and mitigates the risk of catastrophic failures, thus ensuring more stable continuity.

Mimicking and Overlapping Roles

Redundancy also works through *mimicry*. In any organization, the right-hand person often serves as a partial copy of the leader. For example, a development director might closely mirror the CTO's skills and decision-making style, stepping into similar responsibilities when needed. They might mimic the CTO's organizational operating system down to their engineering manager and team layer as well. This redundancy not only reduces the CTO's workload but also ensures that leadership remains coherent and aligned, top to bottom, with the company's vision.

However, mimicry doesn't imply duplication. As the organization grows, individuals naturally develop areas of specialization, adding depth to the system. A development director might focus on operational processes while the CTO shifts to long-term strategy and innovation. This evolving differentiation ensures that redundancy doesn't devolve into waste but instead enhances the system's capacity to adapt and grow.

A CTO might initially feel uneasy or even question their value when they notice the organization running smoothly without their constant involvement. It's natural to associate being "needed" with being indispensable, but this redundancy is actually a sign of a healthy, maturing organization. When a CTO's systems, processes, and team members operate effectively without them, it frees them to focus on higher-level priorities—such as innovation, aligning technology strategy with business goals, and preparing for future growth.

This shift from being in the trenches to steering the ship isn't a loss of importance but a reflection of successful leadership.

Specialization: Differentiation Within Redundancy

Redundancy on its own isn't enough to create strength; its *specialization* that transforms overlap into a powerful advantage. A better analogy can be found in a high-performing sports team: While multiple players may have overlapping skills, each individual also specializes in a specific role that complements the team's overall strategy. Similarly, in business, initial redundancies often evolve into specialized roles over time, allowing the organization to adapt and thrive.

Consider Alice as she recognizes that her growing company needs both visionary leadership and operational precision to succeed. Instead of attempting to handle everything herself, she hires a Chief Operating Officer (COO) to focus on the operational aspects of the business. This operational leader mirrors some of Alice's skills but specializes in managing the internal mechanics of the company, streamlining processes, ensuring cross-departmental coordination, and maintaining operational efficiency.

Meanwhile, Alice shifts her attention outward, dedicating her efforts to cultivating strategic partnerships, expanding market presence, and steering the company's long-term vision. This division of labor enables specialization while maintaining an essential level of redundancy.

If Alice needs to step away temporarily, the COO can seamlessly assume leadership responsibilities, ensuring continuity. Conversely, Alice's foundational understanding of operations allows her to step in effectively if the COO needs additional support. By structuring leadership in this way, the company not only mitigates dependency on a single individual but also becomes more adaptive, with roles tailored to support both external growth and internal stability.

This example reflects how many businesses evolve as they grow. CEOs often transition to focus on visionary leadership while empowering their operational counterparts to manage the complexities of day-to-day execution. This approach demonstrates how redundancy promotes resilience while specialization drives efficiency and organizational effectiveness. However, ensuring that redundancy and specialization work in harmony requires more than structure alone—it demands careful management of the signals that align teams and maintain focus as complexity increases.

Signals in Redundancy and Specialization

Signals play a pivotal role in balancing redundancy and specialization within an organization. Leaders like Alice must amplify signals that drive alignment, such as shared goals, cultural values, and strategic priorities, while suppressing noise that could distract teams from their focus. For example, Alice emphasizes a unified delivery-focused culture across both engineering and sales, creating an alignment of mindset while allowing each team to specialize in respect to execution.

She amplifies this signal through recurring company-wide updates that highlight successful product launches and celebrated client wins. In all-hands meetings, Alice reinforces the message that both timely delivery and customer satisfaction are non-negotiable pillars of success. By doing so, she ensures that teams internalize these shared goals, ensuring coherence across the organization.

At the same time, Alice works to suppress noise that could fragment focus. She actively filters low-impact feature requests from external stakeholders to prevent engineering from being pulled off course. She also aligns with sales leadership to reduce ad hoc client commitments that could strain delivery capacity. By establishing clear communication channels and decision-making processes, Alice shields teams from the reactive urgency that can derail long-term priorities.

This careful regulation of signals allows redundancy to support alignment—such as ensuring both sales and engineering understand the end-to-end customer journey—without stifling the creativity and agility that specialization brings. Engineering can concentrate on shipping reliable features while sales tailors messaging to clients' needs. The shared cultural signal of delivery encourages coherence across teams, ensuring they all align with the company's overarching goals, while their specialized roles maintain decoherence, enabling the organization to remain agile and effectively address diverse challenges.

By managing signals in this way, leaders ensure that redundancy and specialization reinforce one another, creating a resilient system capable of adapting and thriving as the organization scales. Redundancy and specialization aren't opposing forces; they're complementary strengths that fortify the organization. Redundancy builds resilience and safeguards against disruption, while specialization fuels adaptability and precision. Together, they enable scalable, sustainable growth—key attributes of a thriving CAS. For Alice and her team, embracing this balance creates internal alignment, empowers external execution, and sustains the coherence necessary to excel under the Sales sentinel.

Liquid

Chapter 34:
Restoring Flow in a Boiling or Frozen System

When a company finds itself in a boiling or frozen state, overwhelmed by chaos or paralyzed by rigidity, leveraging redundancy and specialization can be a powerful way to bring the organization back into the desired liquid flow state.

Redundancy: Creating Safety Nets to Reduce Chaos

In a boiling system, where chaos and unmanageable complexity dominate, redundancy acts as a stabilizing force. When critical functions are dependent on single individuals or teams, the pressure on those points can exacerbate the boiling effect, with decisions and actions cascading into disarray.

Adding redundancy by overlapping responsibilities, such as assigning multiple people to understand key processes or share workloads, creates a buffer that prevents further escalation.

For example, if a company is struggling to manage simultaneous product launches due to a single project manager's overload, introducing a secondary project lead allows tasks to be distributed more evenly.

This redundancy provides breathing room, reduces bottlenecks, and ensures that no single failure point derails the entire operation.

Redundancy doesn't just replicate roles. It creates resiliency by ensuring the organization can absorb and adapt to pressure points without breaking down.

Specialization: Unlocking Differentiation to Break Rigidity

In a frozen system, where overregulation and rigid structures have stifled progress, specialization can reintroduce the desired adaptability.

Specialization allows teams or individuals to focus deeply on specific aspects of their work, breaking free from the constraints of trying to do everything at once. By delegating tasks according to unique strengths and capabilities, the organization can distribute complexity across multiple specialized units, reducing the burden on any one part.

For instance, in a frozen company where cross-functional teams are bogged down by layers of approval processes, breaking teams into smaller specialized groups can help.

A dedicated compliance team might focus solely on regulatory requirements while the engineering team concentrates on innovation. This specialization clarifies roles, reduces duplicated effort, and restores the organization's ability to move forward effectively.

Increasing Flow: The Liquid State of Redundancy and Specialization

When redundancy and specialization are used together, they create the conditions for a liquid state. A liquid state is one where the system is flexible enough to adapt and structured enough to remain coherent. Redundancy ensures that the company has the resilience to handle disruptions, while specialization allows it to channel efforts efficiently into distinct, high-value areas.

Alice and her team learned this firsthand. After splitting their monoteam into distinct B2C and B2B groups, they introduced specialization, where each group could focus on the parts of the product that were specialized to each customer type.

They further introduced redundancy within their leadership structure by moving Theo to Architect and bringing in Jordan as the CTO. This combination reduced chaotic interdependencies and clarified roles, bringing more effective flow to their organization.

When companies embrace this balance, they can navigate boiling and frozen states more effectively. The effectiveness of a company long-term is directly related to its ability to return to a liquid state where it can grow and evolve while maintaining stability and adaptability.

When a company fosters clear internal communication and coordination, which ensures that teams work cohesively toward shared goals, it is implementing the Sales sentinel more effectively.

This alignment equips the organization to respond nimbly and effectively to external market demands, aligning internal processes with the needs of customers and partners.

PART 5:

Becoming the Catalyst You Need to Be

You have to understand. Most people are not ready to be unplugged. And many of

them are so inured and so hopelessly dependent on the system that they will fight to

protect it.

–Morpheus, *The Matrix*

Liquid

Chapter 35:

Seeing the Secret World

To effectively navigate and influence the complex adaptive system (CAS) within our companies, we need to transform into leaders who understand the system and how to influence it. This requires a deep comprehension of boundaries, agreements, contracts, and the subtle art of influence.

Recognizing the typical patterns of how systems operate allows us to use consistent playbooks, putting standard action plans in place to address specific issues. By seeing the complexity of the system, we can simplify the actions needed to course-correct and guide it toward our desired outcomes.

Understanding and Influencing the System

We must first understand the intricacies of the system. Boundaries define the limits within which various components of the system interact. These can be physical, organizational, or conceptual boundaries that help delineate responsibilities and areas of influence.

Agreements and contracts formalize these boundaries, setting clear expectations and rules for interaction. Influence, however, isn't about strict control. Influence is about guiding the system toward desired behaviors and outcomes. We cannot control the system to bend to our desires; instead, we must nurture and encourage the system to partner with us for success. The system, which is our business, must develop healthy habits in order to succeed. By recognizing standard patterns in how systems operate, leaders can anticipate potential issues and apply proven playbooks to address them. This strategic approach transforms complexity from a source of chaos into a lever for driving clarity, adaptability, and sustained performance.

Acknowledging the Nature of Systems

As we've learned throughout this book, systems will boil and freeze based on the actions or inactions happening within them. These fluctuations are natural: Systems are inherently dynamic, constantly adjusting to new inputs, environmental changes, and internal feedback loops. A system's state, whether it's thriving, stagnant, or declining, reflects the cumulative impact of countless interconnected decisions and interactions within it that have happened over time.

To lead effectively within a system requires us to embrace its fluid nature. Systems respond to influence, but they also resist rigid control. Trying to impose strict, unyielding structures often results in unintended consequences, such as stifled innovation or misaligned priorities.

Leaders who instead focus on consistently nurturing the system by providing the right signals, reinforcing boundaries, and fostering adaptability will encourage the system to evolve positively over time. To grow our company, we need to focus on evolving the entire system operating in our business.

This means continuously monitoring its patterns, gathering feedback, and making iterative adjustments to keep it aligned with the company's goals. Aaron Dignan, in *Brave New Work*,[21] emphasizes that organizations resemble gardens more than machines; they thrive through nurturing, responsiveness, and continuous interaction over time.

Just as a gardener tends to plants by addressing their specific needs—watering, pruning, and creating the right environment—leaders must actively tend to their businesses, ensuring that each part contributes to the overall health and growth of the organization.

Evolving a system also requires recognizing its current state. Is it boiling, frozen, or liquid? Identifying where the system is stuck or overheated enables leaders to take deliberate, targeted actions to restore balance. This isn't a one-time fix; it's a commitment to an ongoing process. Successful systems grow incrementally, building on small, consistent improvements that accumulate over time into significant, sustainable change. By accepting this iterative nature, leaders guide their organizations toward continuous evolution and long-term success.

The Importance of Informed Actions

It's impossible to blindly tweak our way toward success. While some actions may serendipitously influence the system positively, without a fundamental understanding of the system, we're likely to experience numerous failures before stumbling upon a successful strategy.

Also consider that the consequences of actions taken today will most likely not manifest until much later. Firing a team member, adjusting a product road map, or expanding the scope of a solution may all be actions we take to address the perceived hot button issues of the day, but the ripple effect won't be known for months if not years to come.

This trial-and-error approach can be frustrating and inefficient, leading to wasted resources, diminished morale, and misalignment among leaders about the root cause of problems and the best way to address them.

The moment we recognize that our current approach isn't working, we summon the courage to pivot and experiment with new strategies to influence the system. This requires letting go of rigid plans and being open to adaptation, even in the face of uncertainty. True leadership within a CAS means not only acknowledging when adjustments are needed but also having the fortitude to implement changes that may challenge the status quo.

In Part 2, we learned that by seeing the complex system operating in our business, we can make more informed, strategic decisions grounded in the system's dynamics. Instead of reacting to symptoms, we can address root causes, ensuring that our actions align with the broader goals of the organization. This understanding also promotes alignment among the leadership team, enabling them to agree on priorities and work cohesively toward common objectives. Shared insights into the system reduce friction, minimize guesswork, and increase the likelihood of sustained success.

When leaders take informed actions, they empower their teams to navigate complexity with clarity and confidence. This not only accelerates progress but also strengthens trust within the organization, creating a culture where every decision feels purposeful and aligned with the company's mission. Informed action isn't just a leadership skill. Informed action is the cornerstone of effective influence within a CAS.

Liquid

Chapter 36:
Why Leaders Struggle to See the System

One reason leaders struggle to influence the system is that they don't see it in the first place. Instead, they see discrete problems through a simplistic lens: a delayed project, high turnover in a certain department, or declining revenue. Without understanding the underlying interconnections, they attempt to tackle each issue in isolation.

Senge writes in *The Fifth Discipline*[22]: "The fixation on events is actually an affliction of Western culture, which leads us to see the world in snapshots rather than as a process. This is one of the reasons why so many people are unable to see systemic causes—they look at events instead of patterns of behavior."

A CTO might try to improve delivery by hiring more developers, without recognizing that the team is already boiling under the weight of poorly defined processes and overwhelming complexity. More developers simply add to the chaos, making the system boil faster and productivity grind to a halt.

Another common mistake is believing that control is the solution. CEOs might enforce rigid processes or metrics, assuming they will create order. But overly rigid systems will rapidly freeze the business, stifling innovation and adaptability.

The result is frustration, disengagement, and an organization ill-equipped to respond to new challenges.

This isn't an easy journey. Leaders who fail to see the system often find themselves firefighting endlessly, reacting to crises instead of addressing root causes. When systems fly off the gears, when boiling systems implode or frozen systems shatter, it's often the leaders who are blamed.

Leaders get fired not because they didn't act but because they acted without understanding.

Chapter 37:

Seer, Orchestrator, and Believer

Being a Catalyst for an organization requires more than understanding the theory of complex systems. It demands courage, a willingness to challenge our own assumptions, and, most importantly, deep self-awareness. We must understand which roles we're comfortable in and then step into the discomfort of change to proactively guide the organization through transformation.

To be a productive, effective leader within the complex adaptive system (CAS) of our business, we need to embody the roles of a seer, an orchestrator, and a believer:

- As a **seer**, we must develop the vision to perceive the intricate web of interactions and dynamics within our organization. This means recognizing patterns, anticipating changes, and understanding the broader impacts of our decisions.

- As an **orchestrator**, our role is to harmonize these interactions, coordinating efforts across different subsystems to ensure they align with the overall goals of the organization. This involves strategic planning, resource allocation, and promoting collaboration among team members.

- As a **believer**, we need to have unwavering faith in our team's ability to adapt and thrive within the CAS. This belief translates into empowering our team, nurturing their growth, and fostering a culture of transparency, continuous improvement and innovation.

This isn't about perfection. It's about progress. Catalysts influence the system incrementally, making adjustments that move the organization closer to a state of flow. They do this with patience, persistence, and the humility to course-correct when needed. By questioning ourselves as a Catalyst, we commit to a journey of continuous growth and transformation both for ourselves and for our organization.

Am I a Seer?

The role of a seer begins with curiosity and observation. Are you willing to look beyond the surface-level tasks and outcomes and explore the underlying patterns shaping your organization? A seer recognizes that every decision, interaction, and process has ripple effects within the system. This means honing your ability to identify correlations, anticipate phase changes, and discern where the system is thriving or stuck.

Ask yourself:

- "Am I looking for the deeper causes of problems, or am I reacting to symptoms?"
- "Am I cultivating the ability to see the invisible connections and interdependencies that influence my business's success?"

If you aren't actively working to refine this vision, you may be missing critical opportunities to guide your organization toward sustainable growth.

Am I an Orchestrator?

As an orchestrator, you are the conductor of the symphony, ensuring that the individual elements of your organization work together harmoniously. This requires a deep understanding of your company's structure, boundaries, and signals. It also involves making intentional decisions about resource allocation, prioritization, and collaboration. Are you aligning your teams and processes to the company's overarching goals, or are you allowing misaligned efforts to create friction and inefficiencies?

Ask yourself:

- "Am I empowering teams to collaborate effectively while maintaining their autonomy and holding accountability?"

- "Am I ensuring that the signals I send through my decisions and actions are clear, consistent, and constructive?"

The orchestrator must be both strategic and empathetic, creating a framework that enables others to do their best work.

Am I a Believer?

Being a believer means trusting in your team's ability to adapt, grow, and innovate within the complexity of your system. A believer builds a culture of accountability and empowerment, where individuals feel supported to take risks and learn from failure. This belief isn't blind optimism but a deeply held conviction that, with the right guidance and structure, your organization can thrive even in the face of uncertainty. Ask yourself:

- "Do I genuinely believe in the capabilities of my team, and do my actions reflect that belief?"
- "Am I creating an environment of trust and psychological safety where people feel valued and empowered to contribute?"

Without belief, it's impossible to inspire others to embrace the challenges of navigating the complex system that is your business.

Chapter 38:
Balancing Courage and Patience

As leaders navigating the complex system at work in our business, we're called upon to embody the roles of seer, orchestrator, and believer. Yet recognizing patterns, coordinating teams, and holding belief in our vision is not sufficient on its own. These capabilities must be paired with the fortitude to act decisively and the patience to let systemic changes take root. True leadership lies in mastering both.

Do You Have the Fortitude to Make Hard Decisions?

Leadership often demands difficult choices. When we uncover areas of systemic boiling or freezing, it requires tenacity and resolve to address them head-on, especially if it means making unpopular decisions.

For Alice and Theo, this meant splitting their team, investing in new leadership, and fundamentally rethinking their organization.

For you, it might involve reallocating resources, restructuring teams, or letting go of long-standing practices or people that no longer serve the company.

These decisions require conviction—standing firm in the face of resistance and navigating the fear of short-term disruption. Acting decisively might mean breaking established norms or addressing misalignment among leadership, but it's through these bold actions that we guide the system toward growth and stability.

Do You Have the Patience to See It Through?

Equally critical is the patience to allow our actions to unfold. The complex system within our business doesn't change overnight. The interventions we make—like setting new boundaries, encouraging cultural shifts, or introducing key roles—take time to propagate.

These changes require consistent reinforcement to take root as healthy habits within our business. Habits emerge through repeated actions, signals, and feedback loops that align with our broader vision. Our ability to effectively influence the system is directly related to how well our teams build the habits required to remain in the liquid, productive state. Organizational flow is the result of collective habits that align teams, clarify boundaries, and reinforce shared goals.

The temptation to revert to old habits or chase quick fixes can be strong, especially when immediate results aren't visible. It can be particularly challenging when the CEO is pushing the CTO to drive more change. Yet, as leaders, we must hold the course, trusting in the process and the resilience of our team. The signals we send, the boundaries we set, and the cultural shifts we foster act as scaffolding for these new habits.

Over time, they become ingrained behaviors that keep the system flowing: dynamic, adaptable, and aligned. Patience isn't passive. It's an active commitment to monitoring progress, reinforcing positive patterns, addressing negative patterns, and adjusting our influence as needed. Building habits, like cultivating a liquid state, is a gradual process that ultimately creates an ecosystem where collaboration and progress thrive.

Courage and Patience in Action

Being a seer, orchestrator, and believer isn't enough if we're unwilling to act. True leadership comes from balancing the courage to make transformative decisions with the patience to see those decisions through. It means stepping into discomfort, trusting the system's capacity for change, and staying resolute in guiding it toward the ideal flow state.

The CTO role isn't just about perceiving the system's complexity but translating those insights into actions that drive the business forward. This often means working closely with the CEO to more closely align technical priorities with business goals. Imagine a conversation where the CEO shares an ambitious vision, and you, as the CTO, pause to assess its implications for the technology stack, team bandwidth, and timelines. Together, you chart a course that respects both the system's capacity for change and the courage to pursue transformative innovation.

This dialogue embodies the delicate dance of leadership, where we balance vision with pragmatism to steer the company toward its ideal flow state.

The question isn't just whether we can see the system but also whether we can shape it through our actions. Are you ready to take the risks necessary to influence the system effectively? And, just as importantly, are you prepared to nurture and sustain those efforts until the results become visible? Answering "yes" to both is what separates leaders who react to problems from those who catalyze lasting change.

Chapter 39:
Self-Reflection as a Catalyst for Change

Understanding our unique strengths and weaknesses, as well as the areas that bring us the most joy, is critical for CTOs in shaping our roles and guiding the evolution of our teams and organizations.

Every CTO brings a unique blend of strengths to their role. Some excel in crafting strategic visions, others shine in operational execution, and some are driven by a passion for technical innovation.

The CTO Levels framework illustrates how the role of a CTO evolves alongside the company's growth, requiring us to adapt our focus and skills to meet the shifting needs of the business at each new stage—or courageously step to the side (or out) when the two no longer align.

The key to building a resilient and effective organization lies in understanding what type of CTO we are, then using that insight to create a leadership structure that addresses gaps, amplifies strengths, and supports long-term growth both for ourselves and for companies we serve.

Theo's Journey: A Case Study in Self-Reflection

Theo's transition from CTO to Architect marked a pivotal turning point for both himself and the company. As the organization scaled and faced new challenges, Theo took the time to evaluate his role critically.

He reflected on the aspects of the job that energized him, such as crafting scalable technical solutions, defining long-term architectural vision, and ensuring the technical infrastructure could handle future growth.

At the same time, he recognized the aspects of his role that stretched him beyond his comfort zone, such as operational management, scaling processes, and cross-departmental collaboration. Through self-reflection, Theo reframed these challenges not as failings but as opportunities to support the organization's evolution.

This clarity empowered Theo to step into the Architect role and advocate for bringing in a new CTO who could focus on scaling the company's technology strategy and aligning it with broader business goals.

His decision wasn't just about his personal growth; Theo was also a Catalyst for the organization to adapt and thrive.

Hiring to Complement Strengths and Address Weaknesses

Just as it did for Theo, self-reflection can guide how you shape your leadership team. If your strengths lie in strategic thinking, prioritize hiring operational expertise. Consider bringing on a right-hand person, like a VP of Engineering or Director of Operations, to handle process management and team dynamics. If you're deeply technical, you might hire someone skilled in cross-functional collaboration to bridge gaps with other departments.

For Theo, hiring a CTO with a knack for operational scaling ensured the company could continue to grow while continuing to leverage his technical expertise in a focused and impactful way. This shift allowed the leadership team to address gaps without compromising the organization's momentum, and it set a precedent for how the organization could adapt and scale effectively.

Building an Adaptive Organization

An adaptive organization is one that continuously evolves its structure, processes, and leadership capacity to respond effectively to change. Rather than relying on rigid hierarchies or fixed plans, an adaptive organization can sense shifts in its environment, adjust priorities, and redeploy its people and resources to meet emerging challenges.

This adaptability isn't reactive; it's a cultivated capability that allows the organization to maintain resilience and forward momentum amid uncertainty.

The foundation of an adaptive organization is aligning leadership roles to individual strengths, hiring to fill gaps in expertise, and fostering healthy redundancy across critical functions. In the early stages, redundancy often means simply having someone to share the workload. At the same time, hiring specialists addresses capability gaps. Over time, these complementary approaches naturally lead to the differentiation of roles, enabling the organization to develop deep expertise while ensuring no single point of failure.

This evolution results in more than just a stronger leadership team. It produces an organization that can respond with agility to both internal and external pressures. By designing a system that continuously aligns leadership capacity with the organization's needs, leaders cultivate a structure that can adapt to complexity, scale sustainably, and thrive in the face of disruption.

Transforming the Organization

Theo's journey highlights how self-reflection and intentional role-shaping can transform an organization. His decision to focus on architecture didn't isolate him; it amplified his impact. By mentoring the new CTO and leveraging his deep knowledge of the system, Theo helped the company evolve into a more aligned and scalable entity.

It's essential to recognize that this moment of reflection and courage isn't exclusive to the CTO role. Every leader, including the CEO, will one day face the same crossroads—a moment where they need to assess whether their unique strengths and focus are continuing to serve the company's trajectory or not. Embracing this reality is a sign not of weakness but of true leadership that will ensure the business thrives, even if it means stepping aside to make space for the next chapter of its growth.

Self-reflection isn't just a personal exercise. It's a powerful catalyst for organizational change, enabling leaders to influence their teams and systems more effectively. By embracing self-awareness, redundancy, and specialization, you can create a leadership structure that thrives on collaboration, clarity, and collective strength.

Liquid

Chapter 40:

Are You Ready to Be the Catalyst?

The role of a Catalyst is not for the faint of heart. It requires courage, clarity, and a willingness to step into the unknown. To become the leader our organization needs, we must first ask ourselves a fundamental question: **Am I ready to fundamentally change?**

Becoming a Catalyst means stepping away from the comfort of familiar patterns and embracing the humbling realization that much of what we thought we knew about leadership needs to evolve.

Once we see the system at work within our organization, with its hidden correlations, interactions, and emergent behaviors, we can no longer unsee it. The complexity of the system may initially feel overwhelming, and we may feel completely inadequate to be the person to evolve it, but acknowledging it is the first step toward harnessing its power.

Once we can see a problem, we can shift it. But seeing the system is not enough—are you ready to act? Recognizing the complexity within our organization is only the beginning.

A true Catalyst doesn't just observe. A true Catalyst influences. They take the insights gleaned from seeing the system and use them to guide deliberate, thoughtful actions that promote growth, adaptability, and alignment. This requires not only vision but also the fortitude to make hard decisions and the patience to see them through.

Being a Catalyst also means accepting responsibility, not just for successes but also for failures. When systems stagnate or spiral into chaos, the burden often falls on the leader. Are we ready to fully take ownership for the state of our organization, to cultivate trust, and to create the conditions for meaningful change?

This isn't about perfection or control. Remember, we cannot directly control a complex system; we can only influence it. A Catalyst leads by seeing the bigger picture, orchestrating interactions across teams, and believing in the people who make up the system. It requires us to be a steady hand that guides the organization through challenges, the visionary who inspires alignment, and the advocate who empowers others to thrive.

This revelation is often a humbling, even unsettling, experience for most leaders. It forces us to confront the uncomfortable truth: Much of what we thought was effective leadership may have been focused entirely on symptom mitigation rather than resolving root causes. Influence requires consistent action *over time* toward clear goals.

The instinct to control is deeply ingrained in many leaders. Faced with a business that's in chaos or stagnation, they often reach for direct actions to "fix" the problem. They issue mandates, reorganize teams, fire people, or micromanage processes. While these actions might offer a temporary illusion of control, they rarely address the underlying issues.

The complex system at play within our business is not a simplistic machine where flipping a switch or tightening a bolt solves the problem.

It's a living, interconnected network of people, processes, and dynamics that evolves in response to actions—sometimes in unpredictable ways.

Being a Catalyst demands humility. Humility requires the willingness to admit that the system is beyond any one person's control. Humility requires the patience to resist the urge to act out of frustration or fear. True influence comes not from reacting to symptoms but from building trust, aligning teams toward a shared vision, and nurturing the conditions needed for the system to adapt positively over time.

Trust is particularly critical. Effective leaders trust their teams to take ownership within their defined boundaries, and teams must trust their leaders to provide clear, consistent signals and direction. Without this mutual trust, actions are likely to be fragmented and short-lived. A Catalyst fosters this trust by embodying transparency, accountability, and empowerment, encouraging the system to stabilize into a liquid state where adaptability and collaboration thrive.

Above all, a Catalyst understands that the changes they seek are systemic and require sustained effort. Each decision, signal, or intervention nudges the system incrementally. Progress isn't immediate or linear. Progress builds over time.

By embracing this mindset, leaders can move beyond command-and-control tendencies and step into a role of influence. Leaders take ownership to guide their organization through complexity toward sustainable growth and innovation. So, take a moment to reflect: Are you ready to be the Catalyst your organization needs?

The journey will demand your full attention, your best energy, and your willingness to grow. If you commit, the rewards will be extraordinary—not just for your company, but also for the people within it and for you as a leader.

This is your moment to step forward. Will you take it?

Chapter 41:
Using Sentinels as a Measure of Success

As our organization grows and complexity increases, achieving sustainable success requires more than just focusing on individual tasks.

It demands a systems-oriented approach guided by key indicators that help us navigate and influence the underlying system. This is where the Four Sentinels outlined in this book come into play. They serve as the emerging properties within our complex system that guide us toward success. They help us assess the health of our organization and adjust our leadership strategies accordingly.

The Four Sentinels are:

- Speed
- Stretch
- Shield
- Sales

Each sentinel represents a critical aspect of our organization's flow and resilience. Together, they provide a comprehensive lens through which we can evaluate our company's progress and ensure it remains adaptable and thriving.

Speed

Speed is about our organization's ability to deliver quickly and efficiently. It's not merely about working faster but creating the conditions that allow teams to respond swiftly to customer needs, market shifts, and internal challenges. Leaders must prioritize reducing bottlenecks, streamlining workflows, and ensuring teams have the autonomy to move with agility.

Evaluate your current delivery processes. Are there recurring bottlenecks or delays? Where can you reduce complexity and empower teams to accelerate delivery?

Stretch

Stretch signifies our organization's capacity to scale and adapt sustainably. It reflects our team's ability to grow, take on increased complexity, and continuously evolve alongside the business. Stretch requires leaders to invest in developing their people, fostering resilience, and ensuring systems can expand without breaking under pressure.

Assess your team's growth trajectory. Are you equipping individuals to handle greater complexity? How are you preparing your organization to scale effectively?

Shield

Shield represents our organization's ability to protect itself from risks and maintain stability. It encompasses security, compliance, operational resilience, and the safeguarding of our culture. Leaders must proactively identify potential threats and ensure protective mechanisms are integrated into the company's operations.

Review your risk management practices. Are your security measures robust? How prepared is your organization to handle disruptions and maintain stability under pressure?

Sales

Sales isn't just about revenue generation; it reflects our organization's ability to align internally and communicate value externally. It involves selling our vision to our team, aligning cross-functional priorities, and effectively engaging customers. Strong sales capability ensures that our organization remains connected to market needs and can influence internal stakeholders.

Reflect on your organization's internal and external alignment. Are you effectively communicating priorities across teams? How well is your company translating customer insights into business outcomes?

The Four Sentinels aren't independent; they're interconnected. Success emerges when Speed, Stretch, Shield, and Sales work in harmony.

Leaders must continually assess these sentinels, recognizing when one is lagging or overemphasized, and recalibrate their approach to maintain effective organizational flow.

By integrating the Four Sentinels into our leadership mindset, we gain a powerful tool kit for navigating complexity. These sentinels provide clarity amid uncertainty, enabling us to guide our organization toward sustained growth and resilience.

Chapter 42:
Understanding Our Catalyst Role

A catalyst initiates and accelerates change, not through force or direct control but by creating the conditions that allow transformation to happen. In a business context, the Catalyst doesn't impose order or dictate outcomes.

Instead, they understand the complex web of interactions within the organization and influence it in a way that aligns with the company's goals. A Catalyst cultivates adaptability and continuously guides the system toward a healthier, more productive state of flow.

Embracing Influence Over Control

Many leaders are trained to think in terms of control: assigning tasks, enforcing rules, and driving outcomes through direct action. This approach is understandable because tasks are tangible and clearly defined, providing a sense of order.

Humans are naturally inclined to simplify anything that feels complex, but this urge can lead to a superficial understanding of the system and ultimately hinder meaningful progress. But in a complex system like a business, *control is an illusion*. An organization isn't a machine where turning a dial or flipping a switch yields predictable results. Instead, an organization is a living, interconnected, highly complex system where actions in one area create ripple effects across the whole. The Catalyst recognizes and fully accepts this reality and shifts their focus from control to influence. They understand that the only way to lead effectively is to identify the levers that guide the system's natural tendencies and amplify the signals that steer it toward success. This requires patience, observation, and the ability to embrace uncertainty.

A Catalyst doesn't force change. A Catalyst creates the environment where change can happen.

Seeing the Hidden System

A Catalyst sees beyond the surface-level tasks and outcomes and into the deeper system at play. The challenges our organization faces (e.g., missed deadlines, misaligned teams, stagnant innovation) aren't isolated issues. They're symptoms of the system's underlying dynamics at work. Without understanding these dynamics, any solution is merely a Band-Aid, destined to fail when the next symptom emerges.

For example, if product releases are consistently delayed, a reactive leader might focus on adding more developers or enforcing stricter deadlines. These quick fixes often address the symptom but fail to prevent future breakdowns.

In contrast, a proactive Catalyst digs deeper:

- Is the delay caused by poor communication between teams?
- Is technical debt piling up?

- Are there misaligned priorities between product and engineering?

Proactive leaders like Catalysts recognize that solving surface-level problems without addressing systemic issues only leads to recurring crises.

By identifying and addressing the root causes, a proactive Catalyst influences the system itself, creating lasting improvement and ensuring the organization operates in a state of sustainable flow.

Empowering the System to Evolve

The role of the Catalyst isn't to "fix" the system but to empower it to adapt and evolve. The Catalyst creates boundaries that define how teams interact, promotes transparency to ensure everyone has the information they need, and builds trust so individuals feel empowered to make decisions within their scope.

A Catalyst understands that the system's health depends on its ability to self-organize, co-evolve, and respond to new challenges. This isn't a one-time effort. Systems are dynamic, and their needs change over time. A Catalyst remains actively engaged, continually assessing the system's state and making adjustments as needed. They act as a steward of the system to ensure it remains in a productive state of flow even as complexity grows.

Seeing the System Clearly

Catalysts begin by deeply observing the organization and identifying the hidden dynamics at play. They recognize the correlations, dependencies, and emergent behaviors that others might miss.

This process starts with asking the right questions:

- Where are we boiling?

- Where are we frozen?

- Are our teams aligned with the company's goals, or are they working at cross-purposes?

Taking Incremental, Purposeful Steps

The Catalyst understands that change in a complex system cannot be imposed all at once. Instead, they make incremental adjustments, observe how the system responds, and then adapt their approach accordingly. These changes might include introducing new boundaries, refining communication channels, or fostering a culture of transparency.

When faced with a team that's overburdened and falling behind, a Catalyst might focus on redistributing workloads, automating repetitive tasks, or breaking large projects into smaller, more manageable subprojects.

Each action is designed to reduce complexity and create flow without overwhelming the system or the people within it.

Balancing Influence and Autonomy

A key part of being a Catalyst is knowing when to lead directly and when to step back and empower others. Catalysts trust their teams to take ownership of their work while providing the guidance and support needed to keep them aligned with the company's broader goals. This requires a careful balance to amplify signals that matter and suppress noise that distracts.

A Catalyst might empower a development team to experiment with new tools or processes to improve efficiency, while also ensuring that these experiments align with the overall strategic objectives. By striking this balance, the Catalyst encourages innovation as well as maintaining coherence within the organization.

Having the Patience to See Change Through

Catalysts understand that meaningful change takes time. Systems don't adapt overnight, and the path to a liquid state of flow is rarely linear. Catalysts resist the urge to chase quick fixes or revert to old habits when progress feels slow. Catalysts remain committed to the process, continuously monitoring, learning, and iterating.

Introducing a new business operating system, like OKRs, may initially feel cumbersome as teams adjust. A Catalyst remains patient and provides consistent reinforcement and support, knowing that the long-term benefits of alignment and focus outweigh the short-term discomfort of change.

Creating a Lasting Impact

The actions of a Catalyst ripple throughout the organization, creating positive momentum and sustained growth. By addressing systemic issues, encouraging collaboration, and empowering teams, Catalysts help their companies adapt to change and thrive in complex environments.

The work of a Catalyst doesn't end with solving today's problems. A Catalyst builds a resilient organization capable of navigating the unknown challenges of tomorrow. This is the essence of leadership in a complex adaptive system: We don't just react to change, but we actively shape the system to ensure long-term success.

Chapter 43:

Gaining the Knowledge and Seeing the Patterns

To effectively navigate and influence the complex adaptive system (CAS) within our organization, it's essential to develop a deep understanding of the system's complexities and the patterns we can apply to influence it.

There are several types of complexity that can show up within our system. When we can recognize these types of complexity happening, the act of naming them helps us start to learn how to shift them:

- technical complexity
- organizational complexity
- cultural complexity
- customer complexity
- domain complexity
- market complexity
- regulatory complexity

Recognizing and leveraging these patterns allows leaders to anticipate changes, make informed decisions, and implement strategies that promote growth and adaptability.

Knowledge within a CAS encompasses the insights and information that help us comprehend how the system operates.

This includes understanding the interdependencies between different components, the feedback loops that drive behavior, and the emergent properties that arise from these interactions. By accumulating and sharing this knowledge, leaders can build a comprehensive view of the system, enabling more effective management and adaptation.

Recognizing and Leveraging Patterns

Patterns are recurring sequences of events or behaviors that reveal how a system functions. These patterns provide a blueprint for understanding the underlying dynamics of a CAS. By identifying and analyzing these patterns, leaders can predict potential issues and opportunities, allowing for their proactive management.

Recognizing standard patterns in a CAS allows leaders to use consistent playbooks. Playbooks are standard action plans tailored to address specific situations. For example, if a particular pattern of system failure is observed, a predefined set of steps can be implemented to mitigate the issue. This approach streamlines decision-making, reduces reaction time, and enhances the system's resilience.

Seeing the complexity of the system through the lens of patterns allows leaders to simplify the actions they take to course-correct. Instead of reacting to individual symptoms, leaders can address the root causes that drive these patterns.

This strategic approach ensures that actions are more targeted and effective, leading to sustainable improvements within the system.

The Ebb and Flow of Systems

A fundamental acknowledgment about systems is that they will always ebb and flow based on the actions or inaction happening within them. This natural fluctuation requires consistent nurturing and alignment of actions to ensure the system evolves in positive ways. Leaders must be vigilant, continuously observing the system's behavior and adjusting their strategies accordingly.

Incremental, continual growth with a plan is essential for influencing the system to evolve productively. By making small, consistent improvements, leaders can guide the system toward greater efficiency and effectiveness. This approach prevents the system from becoming stagnant or chaotic, ensuring steady progress and adaptation.

It's impossible to blindly stumble your way toward success. While some actions may occasionally produce positive outcomes by chance, a lack of fundamental understanding will more often result in repeated failures before any real progress is made. Without a clear understanding of the system's current state, leaders may misidentify the root causes of problems, leading to conflicting actions that further complicate the situation.

Proactively Planning for Complexity

Proactively developing an action plan for managing complexity is essential to prevent the system from spiraling into crisis. Just as companies create disaster recovery plans to prepare for unforeseen events, organizations need a *complexity action plan* to address the intricate challenges that arise as they grow.

A disaster recovery plan is designed to help a company recover from unforeseen events like natural disasters, data breaches, or system failures. It includes predefined procedures and protocols to ensure the business can quickly resume operations with minimal disruption.

Similarly, a complexity action plan prepares an organization to handle the increasing complexity that comes with growth, ensuring the system remains manageable and resilient.

Humans devolve to their worst selves under stress, and so do companies. Proactively developing a complexity action plan helps us act effectively when the system goes into crisis rather than scrambling to develop a plan under duress.

This plan should include strategies for identifying types of complexity within the system and the areas to examine to combat these phase changes. By observing what's happening and locating parts of the CAS that are locked up (either boiling or frozen), leaders can implement targeted actions to address these issues before they escalate.

The complexity action plan should include:

1. **Identification of complexity types:** Clearly define the types of complexity that have arisen within your system, such as technical debt, communication breakdowns, or process inefficiencies. This helps in recognizing the specific areas that need attention.

2. **Road map development:** Create a phased road map that outlines how the organization will proactively evolve its

systems and processes over time. This road map should anticipate future complexity milestones, identify potential phase change risks, and provide a structured timeline for implementing turnaround and/or preventive measures.

3. **Monitoring and measurement:** Implement tools and processes to continuously monitor the system's health. Establish metrics and indicators that signal when the system is approaching a boiling or freezing point. This could include tracking code complexity, system performance metrics, and team productivity.

4. **Actionable playbooks:** Outline specific actions to take when signs of excessive complexity are detected. These steps might include refactoring code, improving communication channels, or simplifying processes. Having these actions predefined ensures that the team can respond quickly and effectively as complexity arises.

5. **Regular reviews and updates:** Just as a disaster recovery plan is regularly reviewed and updated, the complexity action plan should be revisited periodically to ensure it remains relevant and effective as the organization evolves.

The longer we wait to change, the harder that change will become. For instance, Alice and Theo could have benefited from understanding that their initial rush to develop an MVP would lead to a freezing point. Knowing this, they could have proactively budgeted time for reducing complexity shortly after the MVP's release. This proactive approach would have ensured the system remained in a flow state, allowing the business to continue growing and evolving smoothly rather than getting locked up and grinding to a halt.

The CTO Levels framework serves as an example of a *complexity action plan*. It outlines the key areas of complexity within the CTO's purview that arise as a company grows, provides a method to identify if the system is reaching the boiling or freezing point, and offers playbooks to address these issues. The framework helps organizations move quickly to evolve or create new boundaries between subsystems, thereby reducing complexity and allowing the system to continue growing.

By proactively developing a complexity action plan, organizations can manage complexity effectively, maintaining system stability and ensuring sustainable growth.

This approach prevents the system from reaching a crisis point and allows for continuous improvement and adaptation, much like a well-prepared disaster recovery plan ensures business continuity in the face of unexpected events.

Applying Known Patterns to Manage Complexity

As types of complexity within a system begin to show signs of boiling or freezing, the patterns for how to act to reduce this complexity are quite similar. Our goal as leaders is to identify these phase changes within the system and apply the known patterns to reduce complexity effectively.

For example, collaborative discussions are straightforward and efficient with a small group of two to three people. A group of this size can operate without agendas, rely on verbal communication, make quick decisions, and maintain high alignment. However, as the group expands to seven to nine members, the dynamics change significantly. To maintain effectiveness, these larger groups need structured agendas, written notes, and a formal decision-making process. We should only enter these larger discussions with a very clear plan. This ensures everyone remains aligned on the discussion topics and agrees on the path forward.

When a company evolves into a network of teams, reducing complexity involves standardizing the boundaries between these sub-systems. This is why company-level standard business operating systems, like the Entrepreneurial Operating System® (EOS), are so effective. They help reduce complexity within a business, allowing it to grow faster. EOS provides a structured framework that standardizes how teams collaborate, make decisions, and align their goals.

This reduces misunderstandings, streamlines processes, and ensures that as the company grows, its operations remain efficient and coordinated.

Systems of Systems

Every company operates as a complex system composed of numerous interrelated subsystems, each with its own dynamics, dependencies, and challenges. As the company grows, the complexity of this interconnected ecosystem increases. Recognizing how these subsystems interact, and knowing when one of them is reaching a critical phase change, is essential for maintaining balance and fostering sustainable growth. A phase change might manifest as boiling chaos, where a team is overwhelmed and operating inefficiently, or freezing rigidity, where excessive rules and processes stifle adaptability.

Subsystems within a business often exhibit both boiling and freezing characteristics simultaneously. For instance, an engineering team might be overwhelmed by a flood of urgent requests (boiling) while also struggling under overly rigid documentation or approval processes (freezing). These dual stresses can create a feedback loop, where the chaotic workload and inflexible systems exacerbate each other, impeding the team's ability to function effectively. A CAS operating within a company requires constant nurturing and alignment to evolve positively. *Growth is not a linear process.*

Growth involves iterative adjustments, feedback loops, and continuous monitoring of subsystems to prevent these extreme states.

Standardized frameworks, such as business operating systems, can help regulate inter-team complexity by creating clear boundaries and simplifying decision-making processes. These tools, combined with an awareness of the patterns within the CAS, enable leaders to balance the system, ensuring that each subsystem operates efficiently while aligning with the company's overarching goals.

Leaders can guide their organizations toward a state of flow where subsystems work in harmony, adapting seamlessly to new challenges and opportunities, if they have a broader, holistic view.

This balance is vital not only for achieving sustainable growth but also for maintaining resilience in the face of inevitable changes within the business environment.

Liquid

Chapter 44:

The Catalyst's Responsibility

In *The Matrix*, Neo's transformation isn't complete until he begins to see the code underlying the simulated world. Once he understands the rules and can perceive the interconnectedness of everything around him, he gains the ability to influence the system with precision and power.

The same applies to leadership within your organization. The moment you "see" the complex system at work—the signals, the boundaries, the emergent behaviors—is the moment you step into your true potential as a leader. But seeing the system isn't enough. The challenge is to embrace the responsibility that comes with this awareness. Like Neo, your power lies in your ability to influence the system toward your goals, not to control it.

The system will evolve, whether or not you choose to engage with it. Complex adaptive systems (CAS) are always there, and they're not static. They're in constant motion, responding to the actions and inaction within and around them. As a leader, you have a choice: to passively react to the changes as they come or to actively influence the system, aligning it with your company's goals and your team's aspirations. To be the Catalyst is to step fully into this responsibility. It means recognizing that leadership within a system isn't about exerting control.

Truly effective leadership senses when complexity begins to boil beneath the surface, recognizes the recurring patterns that can restore flow, orchestrates alignment across boundaries, and trusts in the system's potential to evolve and thrive.

Becoming a Catalyst isn't a one-time achievement; it's an ongoing journey of growth, reflection, and action. Leadership within a CAS requires humility—the willingness to admit when something isn't working—and the courage to take thoughtful, strategic action in response.

As a Catalyst, you must accept that every action you take sends ripples through the system, influencing outcomes in ways you might not immediately see. This journey demands the curiosity to explore the unknown, the resilience to navigate setbacks, and a commitment to continuous learning. It's not about arriving at a final destination but about evolving alongside your organization, guiding it through the continual phases of growth, challenge, and transformation.

Chapter 45:
Leading Through Influence and Growth

This journey isn't for the faint of heart. It demands clarity to see the system as it truly is, courage to make the hard decisions that lead to growth, and compassion to support your people through inevitable change. Systems thinking isn't just a tool; it's a mindset that we live every day. It's a commitment to lead with intentionality, to adapt ourselves alongside our company, and to cultivate resilience in the face of complexity.

Think back on what you've learned in this book. The Four Sentinels—Speed, Stretch, Shield, and Sales—are more than metrics; they're the emerging properties of a well-functioning complex adaptive system. The concepts of signals, boundaries, and correlations aren't abstract ideas; they're practical levers for influencing your organization toward a liquid state of flow.

One core responsibility we have as leaders is to serve as coaches and mentors, guiding our teams through the complexities of the business environment. Often, we'll perceive the secret world of the system before our team even knows it's there. The skill development required to start recognizing the system's dynamics is immense, and we cannot expect our teams to grasp its intricacies at the same pace as we do.

Our job as leaders is to role-model great habits—showing how to determine the next best action to positively influence the system—and then pull our teams along for the ride. This role demands far more than managing tasks and processes or issuing top-down directives.

It calls for meeting team members where they are, nurturing their development, and empowering them to grow. By adopting a coaching mindset, leaders create an environment where individuals feel supported and valued, encouraging them to rise to their full potential.

Mentorship also plays a vital role in this process. Sharing our knowledge, experiences, and insights helps team members navigate challenges and seize opportunities for growth. Effective leaders actively listen, offer constructive feedback, and provide their teams with actionable playbooks, equipping them to tackle new challenges without feeling overwhelmed.

This guidance ensures that team members feel prepared and capable rather than being left to flounder as they learn through trial and error.

For example, CTO communities like 7CTOs are ideal havens for the growth-minded CTO seeking to share challenges and level up as Catalysts. Peer communities offer a dynamic reflection of the complex systems we lead, providing real-time insights, pattern recognition, and collective problem-solving, all of which accelerate our ability to see the system at work within our organization.

This approach builds a stronger, more resilient team and drives innovation and continuous improvement throughout the organization.

By investing in the personal and professional growth of team members, leaders ensure their organizations remain adaptable and thrive in an ever-changing, complex business landscape.

Chapter 46:
Embracing Failure Leads to Growth

Being willing to acknowledge our own faults and failures is a critical part of being the Catalyst. Imagine a world where you crave feedback because it reveals how the complex adaptive system operates. You're a participant inside this system. The more you understand its dynamics, the more effective you become in influencing it toward flow.

This requires humility. We must admit when we make mistakes and view those mistakes as opportunities for learning and growth rather than reasons for blame or punishment. Failure becomes information— data that reveals how the system is functioning (or not). The patterns we uncover through setbacks are the same patterns that point us toward progress.

When failure is treated as a catalyst for change, innovation and experimentation flourish. This mindset creates a culture where team members are unafraid to take calculated risks. Learning from setbacks becomes essential for managing complexity and sustaining flow in a growing system.

As the system evolves productively and positively, the business grows and evolves alongside it. You know you're moving in an effective direction when the Four Sentinels—Speed, Stretch, Shield, and Sales—emerge consistently as the natural outcomes of this evolution.

This evolution isn't driven by monumental shifts but by continual incremental improvements. Each small adjustment builds upon the last, creating a cumulative effect that propels the organization forward.

Leaders must remain attentive to the feedback loops within the organization, constantly sensing, interpreting, and adjusting as needed. This requires understanding the intricate web of interactions and dependencies within the system, recognizing that changes in one area ripple across the whole.

By embracing this mindset and fostering trust, transparency, and collaboration, you can guide your organization toward sustainable growth and innovation. You don't merely lead the business; you shape the system that allows your business to thrive. Failure becomes an essential input to the system's evolution, and your role as a Catalyst is to ensure your team learns to read these signals, adapt, and continue moving forward.

Chapter 47:
The Future Belongs to Those Who See

The journey of leadership within a complex system doesn't end. It's a continuous process of seeing, sensing, and influencing. As your organization evolves, so too must you. Your leadership isn't static; it's adaptive and ever-changing alongside the system you guide.

Throughout this book, we've explored the ways in which systems can boil with chaos, freeze into rigidity, or achieve a liquid state of flow. Your role as a leader is to sense these states, recognize the shifts between them, and guide your organization toward flow. Flow is where creativity, alignment, and performance thrive. It's where the Sentinels—Speed, Stretch, Shield, and Sales—emerge naturally, signaling a system in motion.

You've begun to see the patterns, the signals, and the states of boiling, frozen, and liquid. You've learned to influence flow rather than force control. You've embraced failure as feedback and seen how small adjustments can create forward momentum. This is only the beginning.

The future of your leadership will be shaped by your willingness to stay present, curious, and humble. Systems leadership isn't about reaching a final state.

It's about leading in the constant state of becoming—where both you and the system grow and evolve together.

As you continue this work, ask yourself:

- How will you continue sharpening your ability to see the system?

- What practices will you put in place to better sense the signals of flow and friction?

- How will you nurture the Sentinels—Speed, Stretch, Shield, and Sales—as emergent outcomes of a thriving system?

- How will you recognize when your system is boiling with friction or freezing into rigidity, and what actions will you take to restore flow?

- How will you show up as a Catalyst when uncertainty clouds the path forward?

The future is uncertain. The system will continue to change. What is certain is that your influence matters. Your ability to see, sense, and act will shape not only your organization's success but also your legacy as a leader.

As this book comes to a close, the real journey begins. The question isn't whether your company will evolve—it will. Instead, consider the questions below:

- In what direction will it evolve?

- Will you evolve with it?

- Will you rise to the challenge, embrace systems thinking, and become the Catalyst your organization needs?

The future of your company is being shaped every day. Now is the time to step up, lead with clarity, courage, and compassion, and guide your system toward its greatest potential.

The path is ongoing. Your leadership journey is just getting started.

Liquid

We Would Love to Hear What You Think

If *Liquid* sparked new ways of thinking, leading, or acting, we'd be grateful if you shared your thoughts. Your review helps others discover the book and adds momentum to the shift toward more adaptive organizations.

Visit your favorite bookseller, scan the QR code, or visit ctosentinel.com/liquid to leave a review.

Liquid

References

1. Uhl-Bien et al., Complexity Leadership Theory: Shifting leadership from the industrial age to the knowledge era, *The Leadership Quarterly*. 2007, 18(4), 298-318. https://www.sciencedirect.com/science/article/abs/pii/S1048 984307000689

2. https://systemsthinkingalliance.org/

3. Etienne de Bruin, https://7ctos.com/

4. Kathy Keating, Etienne de Bruin, Scott J. Graves, https://www.ctolevels.com/

5. Charles Darwin, *On the Origin of Species* (John Murray, 1859)

6. Scott J. Graves, https://www.scaletech.consulting/

7. Kathy Keating, https://kathkeating.com/

8. Donella H. Meadows, *Thinking in Systems* (Chelsea Green Publishing, 2008)

9. Peter M. Senge, *The Fifth Discipline* (Doubleday, 2006)

10. John H Holland, *Emergence: From Chaos to Order* (Addison-Wesley, 1998)

11. Wachowski, L., & Wachowski, L. (1999). *The Matrix*. Warner Bros.

12. Daniel Kahneman, *Thinking, Fast and Slow* (Farrar, Straus and Giroux, 2013)

13. John Gall, *Systemantics* (Quadrangle, 1977)

14. Frederick Brooks Jr., *The Mythical Man-Month* (Addison-Wesley Professional, 1982)

15. John Gall, *Systemantics* (Quadrangle, 1977)

16. *Star Trek: The Original Series*, created by Gene Roddenberry (Desilu Productions, 1966–1969)

17. Matthew Skelton and Manuel Pais, *Team Topologies* (IT Revolution Press, 2019)

18. John Doerr, *Measure What Matters* (Portfolio/Penguin, 2018)

19. https://kathkeating.com/get-your-okrs-out-of-my-gems/

20. https://www.eosworldwide.com/

21. Aaron Dignan, *Brave New Work* (Portfolio, 2019)

22. Senge, P. (1990). *The Fifth Discipline: The Art and Practice of the Learning Organization*. New York: Doubleday/Currency.

About the Authors

Kathy Keating, Etienne de Bruin, and Scott Graves are experienced CTOs and co-founders of CTO Levels, a groundbreaking framework that helps technology executives navigate complexity as their companies scale. Together, they have led engineering organizations from early-stage startups to high-growth enterprises, founded successful companies, and advised hundreds of CTOs and CEOs worldwide.

Kathy is a board-certified director, strategic advisor and CTO coach known for applying systems thinking to cultivate thriving product engineering organizations that deliver transformational growth.

Etienne is the founder of 7CTOs, a global peer network for technology leaders, and an executive coach focused on helping CTOs apply systems thinking to leadership.

Scott is a fractional CTO for scaling startups, specializing in helping fast-growing companies navigate complexity and thrive.

They share a belief: *Technology leaders succeed when they can see the hidden systems at work and become the catalyst for clarity, momentum, and lasting impact.*